Soil Microbiology and Soil Biotechnology

Volume 1

by
Prof. Dr. S. Kannaiyan
Former Vice-Chancellor
TNAU, Coimbatore - 641 003
and
Former Chairman
National Biodiversity Authority (GoI)
Chennai - 600 041

2016
Associated Publishing Company®
A Division of
Astral International Pvt. Ltd.
New Delhi – 110 002

ISBN 9789351306290 (SET)

Published by : **Associated Publishing Company®**
A Division of
Astral International Pvt. Ltd.
– ISO 9001:2015 Certified Company –
4736/23, Ansari Road, Darya Ganj
New Delhi-110 002
Ph. 011-43549197, 23278134
E-mail: info@astralint.com
Website: www.astralint.com

Digitally Printed at : Replika Press Pvt. Ltd.

Dedicated

To

My Beloved

Wife

Mrs. Banumathi Kannaiyan

Contents

(Vol. 1)

Preface

The characteristics of the soil environment vary with climate. Soils differ in depth, physical properties, chemical composition and origin. There are five major categories of soil constitnents such as mineral particles, organic residue, water, gases and biological system. The physical structure, aeration, water holding capacity and awailability of nutrients are determined by the proportion of the mineral particles which are formed by the weathering of rock and he degcadative metabolic activities of microorganisms. The vast differences in the composition of soils together with differences in their physical characteristis and agricultural practices by which they are cultivated, reflect equally large differences in their microbial population.

Variations in climatic conditions may selectively favour certain physiological types. Interactions between and among microbiol species no doubt have an important effect on the members of the population. This situation is an extemely complex conditions. Predatory protozoa and antibiotic producing actinomycetes may eliminate a segment of the flora whereas cellulolytic and proteolytic organisms may provide nutrients for less versatile biochemical species. Plant roots penetrate the soil, breaking up mineral particles, removing nutrients, converting them to organic compounds and sereting metabolic products. When plants and other soil inhabitants die, their organic components are mineraliged by microorganisms, which likewise contribute metabolic excretions and dead cells. In the soil there is a constant and complex cycling of organic and inorganic materials. The most important role of soil microorganisms is their function as biogeochemical agents for the mineralization of organic carbon, nitrogen, phosphorus and sulphur compounds so that these biologically important elements are available for reuse.

Soil Microbiology and Soil Biotechnology covers in a major way on Soil Bio-resources, Soil organic matterand organic manure decomposition, Nitrification and Denitrification process in Soil, Sulphur bacteria and the environment. It also covers Rhizosphere Microflora and Plant Microbe interaction, Microbial Ecology of Arid Soil, Industrial potentials of Microbial pigments, Beneficial applications of the fungal antagonists - *Trichoderma* and Microrganisms and Environment. It also covers Biological fertilizers, *Azospirillum* in Soil and its colonization in plant roots, Liquid Bioinoculants, Mineral Phosphate Solubilization by blue green algae, AM fungi, Role of Host genes in Symbiotic Nitrogen Fixation, Bioecology and *Biotechnological* applications of *Trichoderma,* Transformation Technology against Pests and Diseases, Production Technology of Paddy Straw Mushroom, Oyster Mushroom and Biodiversity of Oyster, Paddy Straw and Milky Mushroom. The Book Soil Microbiology and Biotechnology Volume 1 and 2 covers both basic aspects and applied aspects of Soil Microorganisms.

I am grateful to the contributors who readily agreed to share their rich experience and knowledge in their respective fields of specialization to the important volumes meant for scholars and P.G. students. I profoundly thank M/s Associated Publishing Company, New Delhi and especially its owner and publisher Shri S.K. Dutta for all his sincere efforts in bringing out the volumes expeditiously.

I wish to place on recovd my special thanks to my beloved wife Mrs. Banumathi Kannaiyan, my son Dr. Lenin Kannaiyan, Ph.D. (UK) and my daughter-in-law Dr. (Mrs) Curie Pradeep, M.S. (USA) and my son-in-law Dr. (Mrs) Hemavathy Lenin and my daughter Er. D. Pradeep, M.S. (USA), M.S. (Food Engg. USA) who are always with me by extending their full support and cooperation in all my academic and scientific endeavours.

Thanjavur **Prof. Dr. S. Kannaiyan**

November, 7, 2009

Soil Microbiology and Soil Biotechnology : Vol. 1
Editor : S. Kannaiyan
Associated Publishing Co., New Delhi, 2010, pp. 1-8

1

Soil Bio-Resources

*S. Kannaiyan**

Introduction

Soil—A critical natural resource

- Among the natural resources soil is a dynamic living natural body that is vital to the functions of terrestrial ecosystems.
- Soil serves as a medium for plant growth and oxygen supply for root development.
- The physical, chemical and biological properties of the soil constitute its health towards agricultural use.
- Soil health is inseparable from issues of sustainability and it is enhanced by efficient management techniques.
- More than 5.5 billion people in the world are now using about 10% of the land area of the earth to raise crops and livestocks.

Management of soil—A key component in sustainable agriculture

- Soil degradation is a major constraint for productivity growth in agriculture.
- Rather, soil degradation is a major threat to food security and sustainability.
    - Already, an estimated area of 2 billion ha are degraded, world-wide.
    - Around 5-7 million ha of aerable land are being lost annually.
    - Overgrazing accounts for an estimated 35% of total soil degradation.
    - Another 30% resulting from deforestation.
    - 28% resulting from improper agricultural activities.
    - In Asia, and Africa alone, the combined annual losses has been estimated over 15 billion US dollars, due to soil degradation.

Soil and water conservation

- Out of 329 million ha of geographical area in India 58 million ha are marginal lands with more than 5 tonnes per ha soil loss annually which must be brought under soil and water conservation and restored in sustainable agriculture.

* Former Chairman, National Biodiversity Authority, Neelankarai, Chennai - 600041, Tamil Nadu

- Of the 66.73 million ha area under forests almost 36.73 million ha are badly degraded forest areas which can be brought under soil conservation and reforestation programmes.
- In addition, another 21 million ha are waterlogged, saline and sandy soils.
- 29 million ha of improperly used marginal lands which need to be restored to normal productivity through soil amelioration techniques.

Integrated plant nutrient management system

- Increasing sustainability concerns require a system of nutrient management that makes optimum use of fertilizers and efficient recycling of nutrients from soil resources.
- In this context practising Integrated Plant Nutrient Management, (IPNM) is necessary.
- In India, there are 514 Soil Testing Laboratories which serves to 105 million farm holdings.
- Results of many field trials have shown that overall fertilizer recommendations are only adhoc adjustments and need to be replaced by precise and meaningful soil test based recommendations.
    - For maximizing crop yields
    - Optimizing the input use maintenance of soil fertility
    - Improving agronomic efficiency and
    - Environmental quality
- There is a vast potential and tremendous prospects of increasing the crop productivity, through increased fertilizer use efficiency and the IPNM.

Organic sources in nutrient management

- Organic sources of plant nutrients are to be encouraged for soil health management and sustainability.
- However, the organic sources of plant nutrients either individually or in combination could not susbstitute entirely the chemical fertilizers and only partial substitution or supplementation could be made with the organic sources.
- A judicious mix of one or more organic sources like animal manure legumes in rotation to green manures crop residues or farm organic waste and mineral bearing fertilizers in an integrated manner could sustain the crop production and soil fertility levels on a long term basis.
- The availability of principal agricultural residues also depends upon the area devoted to cereals such as wheat, rice, barley, maize, *Sorghum* and pearl millet and residues of pulses and oil seeds.
- Based on the grain/straw ratio approximately 335 million tonnes of agricultural crop residues are available in India.

Integrated farming system (IFS)

- Development of integrated farming system for different ecosystems is necessary for comprehensive agricultural development.

- In lowland IFS comprising cropping, poultry, fishery and mushroom enterprises focus to recycle the residue and byproducts of one component for the other.
- Bearing in mind the advantage of recycling of cow dung, abiogas unit can be linked to the waste from animal component in the IFS of upland with supplemental irrigation instead of directly utilizing it as FYM.
- The IFS model integrating goat farming as a component with crop activity in rain fed condition would not only increase the income of the farmer but also maintain soil fertility and sustainability.
- There is a vast potential and tremendous prospects of increasing the productivity of crop in lowland/upland/dryland areas coupled with INM and IFS.
- Science has served as a powerful tool for increasing food and agricultural productivity.
- However, our current scientific emphasis on productivity per unit area alone is not sufficient enough for self sustainability.
- Maintaining soil fertility by the application of organic manures and organic crop residues in combination with fertilizers would be the best way for achieving better crop productivity and soil sustainability.
- The most important components which will have greater impact in achieving the ultimate goal of sustainable agricultural development are:
  - Policy on fertilizer use and its management
  - Soils and soil testing by modernizing Soil Testing Lab.
  - Integrated plant nutrient system (IPNS)
  - Integrated farming system (IFS)
  - Development of dryland agricultural areas
  - Integration of leguminous crops and rotation
  - Crop residue management
  - Use of organic wastes and organic manures
  - Enrichment of FYM
  - Use of biological fertilizers.
- There is an urgent need for environment friendly or ecofriendly economically areable but sustainable technologies to satisfy the requirements of huge population in the country.
- Sustainable agriculture requires the management of resources in a way to fulfill changing human needs without damaging the quality of environment and conserving the vital natural resources.

Biofertilizer technology

- Biofertilizers technology enables the use of living organisms for the nourishment of plants, either by fixation of atmospheric nitrogen through the process of BNF or by solubilization of mineral nutrients like phosphorus by phosphobacteria.
- BNF is an important biological process to sustain agricultural productivity and application

of biofertilizers to legume crops and also to other crops is the viable alternative for better crop production and soil sustainability.

- A global inventory of the process of nitrogen supply for crop production indicates that BNF is predominant and approximately 175 million metric tonnes of nitrogen are fixed annually which includes 85 million tonnes fixed biologically in soils.
- This estimate includes 35 million tonnes by legumes, 10 million tonnes by non-legumes and 40 million tonnes by pastures and grasslands.
- The global value of BNF runs to several million dollars.
- It has been suggested that around 200 million metric tonnes of nitrogen is turned over on the earth each year.
- Biofertilizers are apparently renewable, cost effective, ecofriendly, farmers friendly and sustainable.
- The potential biological systems such *Rhizobium, Azorhizobium*, - a stem nodulating bacterium, BGA, *Azolla* are crop specific. It has shown nodulation of xylem colonization of *Azorhizobium* in rice and wheat.
- Biofertilizers like *Azotobacter, Azospirillum,* Phosphobacteria and VAM-Vesicular Arbuscular Mycorrhiza could be regarded as broad spectrum biofertilizers particularly *Azospirillum*—a versatile bacterium in colonizing roots of cereals, rice, maize, *sorghum,*pearl millets and oil seeds *etc.*
- These potential biological fertilizers could be better exploited for crop production in sustainable agriculture.

Vermi-composting

- The traditional technology of using earthworms to eat the garbage and convert it into valuable organic manure namely vermi-compost is gaining considerable importance in maintaining soil fertility and sustainability.
- There are over 3200 species of earthworms in the world and nearly 509 species of earthworms are available in India but only six species are being used for composting the organic waste materials.
- In India, about 700 million tonnes of organic wastes are generated every year and these organic wastes consists of farm residues, rural and urban wastes.
- Many factors have contributed to the renewed interest in earthworms.
- Increased pressures on farmers to reduce the cost of cultivation due to increasing cost of fuel for fertilizers, labour, farm machinery and falling prices of agricultural products.
- Potential hazards to wild life domestic animals and human health due to uptake of pesticides by crop plants and the resultant pollution of the environment.
- An urgent need to optimize nutrient cycling in developing countries where lack of material resources limits reliance on fertilizers and other agricultural chemicals.
- Developing alternative methods that do not lead to excessive losses of soil especially by water erosion.

- Traditional methods of disposal of industrial and domestic wastes leading to heavy metal contamination of the environment.
- Also community interest in the nature conservation and environmental protection.

Coconut coir pith composting

- Millions of tonnes of coir-pith is available every year causing serious pollution.
- It is a complex molecule resistant to microbial degradation and contains lignin 36%, cellulose 26%, and organic carbon 24% N, P, K, Ca, Mg, Fe with a carbon to Nitrogen (C/N) ratio of 112:1.
- Coir pith is not easily decomposed due to the presence of 8 to 12% tanin related to phenolic compounds identified as plant/microbial inhibition.
- The mushroom fungus *Pleurotus* sp. releases a variety of biopolymer degrading enzymes along with laccase enzymes and the fungus is widely used commercially on lignin rich waste to hasten the process of composting. For composting one tonne of raw coir-pith, 5 kg urea and 5 bottles of *Pleurotus* spawn are used.
- The composted coconut coir pith is not only an excellent source of organic fertilizer for annual and plantation crops but also act as an excellent source as organic mulch for crop production.

Phospho-compost

- India is producing 0.4% of world's rock phosphate which is of inferior quality and not used for fertilizer production and 5% of P requirement is by imported rock phosphate.
- India has 260 million tonnes of low grade rock phosphate.
- In phospho-compost the microbial inoculants such as *Bacillus subtilis var. phosphaticum, Aspergillus avamari, Thiobacillus thioxidan* are used with rock phosphate organic waste FYM *etc* for enhancing the process of phospho composting and thereby the available 'P' content in the compost is increased.
- The phospho compost is used as the phosphate fertilizer and superior to super phosphate as it serves good source of P to crops.

Biodigested slurry

- Among the energy related technologies biogas technology has gained momentum because of its efficiency, viability, ready acceptability and easy availability of raw materials in rural areas.
- Bio-digestion is a multi-step complex process involving mixed bacterial population.
- Biodigested slurry is a valuable source of organic manure and it contains 1.4 to 1.8% nitrogen, 1.1 to 2% phosphorus and 0.8 to 1.2% potash.
- It has been estimated to yield 236 million tonnes of manure per year.
- The manure is equivalent to 3.5 million tonnes of nitrogen which is an excellent organic source for crop production and maintaining soil fertility.

Water quality criteria for irrigation

- In irrigated agriculture the hazards of soil water is a constant threat to soil problems.
- If the amount of water leaching through the soil is too low due to low rainfall to remove the salts the soil's salt content increases and crop yield decreases.
- Wastewater recycling has attracted much attention currently to prevent toxic contaminants to reach the regular cultivable land and also safeguard the land and environment.
    - Proper care should be taken before using the different kinds of water for irrigation.
    - Use of sewage effluent water for irrigation.
    - Use of industrial effluents for agriculture.
    - Use of wastewater for land application.
    - Tannery industrial wastewater for agriculture.
    - Paper and pulp waste waters for irrigation.
    - Distillery effluent for agriculture.
    - Sago industry effluent for agriculture.
    - Other Industrial wastewaters.

Bio-remediation

- It is a new approach in wastewater treatment to utilize the treated water for irrigation.
- It is a strategy or process that uses micro-organisms, plants or microbial enzymes or plant enzymes to detoxify the contaminants in soil and other environments.
- The concept includes biodegradation, mineralization and co-metabolism.
- Successful bio-remediation depends on the availability of appropriate microorganisms.
- Waste recycling can bring tremendous benefits to agriculture and land management in the long run.
- In addition there are benefits of cleaner environment, a healthy habitat and an intelligent use of all available recyclable resources without condemning them as wastes.
- Proper recycling produces a number of gains such as supplying essential plant nutrients, improving soil physical properties, reducing their accumulation at or near the sites of production and reducing health hazards and enhancing environmental quality.
- Soil is a living and breathing system which can be chocked or damaged if its capacity to absorb and retain water to keep a balance in air/water relations and the needs and limitations of the diverse organisms which live in it are lost sight off.

Soil and environment

- Soil and water degradation are important aspects of environmental pollution.
- Atmospheric concentration of carbon dioxide and other gases have been increasing steadily.
- It has been reported that terrestrial ecosystems play a key role in the global carbon budget.

- Soil misuse and over exploitation causing rapid decline in organic matter content might affect the emission of green house gases into the atmosphere.
- A reduction of 1% of organic carbon content of soil can lead to an annual emission of about 128 billion tonnes of carbon into the atmosphere.
- Reversing the trends of soil and water degradation is crucial for sustainability.
- The degraded land surface is more in Asia and Africa.
- Only 37% of Indian land area can be said to be free from degradation of any kind.

Approaches for natural resource management

- The need for stakeholder analysis especially the weaker stakeholders.
- Recognition of the indigenous knowledge in the process of finalizing any specific treatment as part of the programme.
- Ensuring stakeholder participation in the planning and implementation process for their co-operation.
- Taking up agricultural intensification in resource endowed areas by sustainable blending of the modern technologies with the indigenous technologies to ensure sustainable production.
- Taking up land improvement and soil and water conservation programmes before going for agricultural extensification in marginal or dry lands to augment production.

Sustainable utilization of resources

- The resource management strategy for sustainable utilization of resources is very important.
- Control of soil erosion and soil degradation.
- Reclamation of saline alkaline and acid soils.
- Management of waterlogged soils.
- Soil conservation techniques in different agro-eco systems.
- Soil testing and soil health monitoring.
- Wasteland development.
- Watershed management, check dams, contour trenches and afforestation.
- INM and IFS.
- Efficient use of nutrient sources.
- Nutrient management in farming systems.
- Micronutrients application in micronutrient deficient areas.
- Crop improvement to match the conditions of problem soils, sodic soils and coastal saline soils.
- Soil utilization according to land capability.
- Land capability classification.
- Use of computer technology–Geographic Information Systems (GIS), Decision Support System, for agrotechnology transfer crop models.

- Water conservation–Integrated watershed management–*in situ* and *ex situ,* rain water harvesting-Scientific irrigation management.
- Conjunctive use of surface and groundwater, fresh and recycled water.
- Micro-level agro-climatic analysis for better planning of agricultural production systems.
- Growing multipurpose trees–Agro-forestry for a sustainable land management system.
- Growing N_2 fixing trees in the field border to reduce soil erosion and also to utilize the leaves for organic matter purpose in soil.
- Growing cereal fodder grasses in Dryland agriculture system considerably reduces the loss of top soil.
- Growing leguminous creepers in soil would bind the soil and significantly reduce the soil loss by acting as a **"Bio Mulch"**.

Conclusion

- In a developing country like India, land is not only an important factor of production but also the means of subsistence for majority of the people.
- Agriculture contributes about 26% to India's Gross Domestic Product (GDP) but absorbs nearly 65% of the country's working population.
- Generally the major constraints to rural development are the low level of agricultural productivity coupled with inadequate farm and non-farm diversification, low level of technological change and lack of infrastructural facilities of irrigation, roads, markets, storage and processing.
- But, high rates of soil erosion, land degradation and ground water depletion in many regions due to indiscriminate use of land, water and other natural resources have passed additional threat to ecological balance and sustainability of livelihood system of the people.
- All types of lands are not equally suitable for profitable alternative enterprises. Hence, cluster approach for development based on agro-climatic as well as techno-economic potentials of each region would be very essential.
- Soil is the basic unit or basic foundation for crop cultivation/crop production.
- Safeguarding and maintaining the soil in a proper way is very important for crop productivity and sustainability.
- Soil is a living biological system and maintaining soil fertility would ensure the productivity to sustain our future generation.

Soil Microbiology and Soil Biotechnology : Vol. 1
Editor : S. Kannaiyan
Associated Publishing Co., New Delhi, 2010, pp. 9-30

2

Soil Organic Matter and Organic Manure Decomposition

D. Balachandar[1] *and S. Kannaiyan*[2]

Introduction

Soil is three-dimensional and highly dynamic. It contains abundant plant and animal life. There are four main components of soil: mineral matter, organic matter, air and water (Fig. 1). The mineral matter is made of sand, silt and clay size particles—the basic texture of the soil. The soil water contains dissolved minerals and is the main source of water and nutrients for plants. The air in the soil is needed for plant roots and soil microorganisms to obtain oxygen. The soil organic matter (SOM) is composed of decomposing residues, by-products formed by decomposition and resistant soil humic materials. Different ecosystems have different amounts of soil organic matter, partly because of different temperature and decomposition rate. Soil organic matter is formed by the decomposition of plant components into low molecular weight components that then repolymerize. The carbon in soil organic matter comes from carbohydrates, which are primarily of microbial origin, once decomposition has been completed or from nitrogen containing compounds, aliphatic fatty acids, alkanes, oils and aromatic compounds come from the decomposition of lignins. Thus the exact nature of soil organic matter is not clear. It has no definite shape or chemical formulae or structure (Coyne, 1999).

Dead animal and plant matter begin to decompose as soon as they are added to the soil. Earthworms, beetles, springtails and collembola, the macro (large) to meso (medium) fauna, begin to break large pieces of debris into smaller pieces. At the same time, the microbial population increases rapidly. The microorganisms consume the animal and plant remains and then die, adding themselves to the organic matter. Some organic matter is more readily decomposed than others. The end product of decomposition is *humus* - dark brown or black organic matter that is highly resistant to further decomposition. Most soil organic matter is present near the soil surface, rather than deeper in the soil. Plant roots become organic matter as their cells slough off or as they die. Aboveground parts of the plant that are not harvested are left on the soil surface when it dies. Earthworms and insects incorporate the surface residues deeper into the soil, but the greatest concentration of organic matter still remains in the top six inches of soil (Copperband, 2002).

1. Department of Agricultural Microbiology, Tamil Nadu Agricultural University, Coimbatore – 641 003, India
2. National Biodiversity Authority, Neelankarai, Chennai – 600 041, India

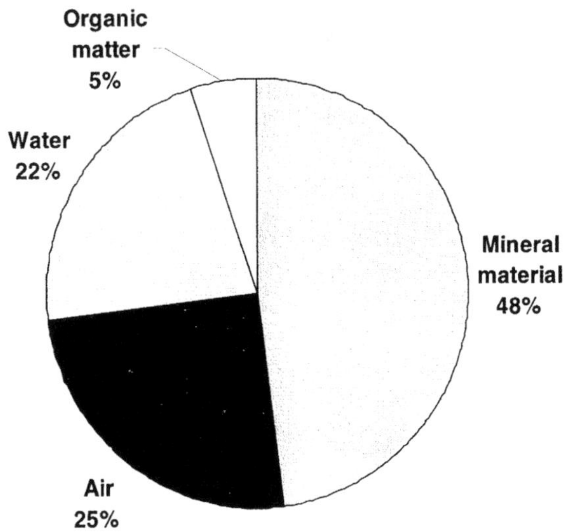

Fig. 1: Four major components and their proportion in soil

Source of SOM

Not all soil organic matter is created equal. Crop residue is a convenient and valuable source of organic matter. Maize, rice, wheat, sugarcane after-harvest will provide high amount of surface residue and root biomass per acre. Dense, sod-type crops like cereals produce generous amounts of root biomass. Soybeans, mung bean like pulses generate very little surface residue. Surface residue encourages the decomposers especially fungi and increases food web complexity. Fungi increase because they have some advantage over bacteria in digesting surface residue, though greener and younger plant residues are easy for bacteria to use. Residue provides food and habitat for surface feeders (some earthworms) and surface dwellers (some arthropods). The rate of decomposition and accumulation of SOM from surface residues depend on the chemical composition of the wastes (Martin and Haider, 1971). Fruit and vegetable wastes are easily degraded because they contain mostly simple carbohydrates (sugars and starches). In contrast, leaves, stems, nutshells, bark and trees decompose more slowly because they contain cellulose, hemicellulose and lignin. The ease with which compounds degrade is determined by the complexity of the carbon compounds and generally follows the order: carbohydrates > hemicellulose > cellulose > chitin > lignin. In contrast to fresh plant residues, composted organic materials decompose slowly when added to soil because they have already undergone a significant amount of decomposition during the composting process. Thus soil organic matter can be differentiated into pools or fractions based on their availability for microbial breakdown (Paul and Clark, 1996). The active fraction has the highest turnover rate of 1 to 2 years. The intermediate pool turns over in two to five years. The recalcitrant or stable pool is well-decomposed organic matter that is chemically or physically resistant to breakdown, taking more than 10 years (Monreal *et al.,* 1997). Well-decomposed organic matter no longer provides as many nutrients for plants and soil microbes as the active pool, but it does play important roles in the soil, such as promoting water and nutrient retention and prevent soil compaction and crusting (Table 1).

Table 1: Estimates of turnover time for soil organic matter in different fractions and soil aggregates (Monreal *et al.*, 1997)

Type of organic matter	Estimated turn over time (years)
Organic matter in fractions	
Crop residues, litter	0.5 – 2
Microbial biomass	0.1 – 0.4
Macroorganic matter	1-8
Light fraction	1-15
Organic matter in aggregates	
Non-aggregated soil	1-7
Macroaggregates (>0.2 mm)	1-23
Microaggregates (<0.2 mm)	3-80
Silt and clay	5-1000

Functions of SOM

SOM is vital to the sustainable use of soil because of its role in maintaining soil structure, water holding capacity, source of microbial biomass, soil fauna and in nutrients cycling.

The maintenance of soil organic matter through rotational farming, effective utilization of crop residues and manures are central to organic farming (Goulding *et al.*, 2001). The soil organic matter affects the retention, release and availability of plant nutrients. It releases immobilized N, P and S during decomposition. Plants obtain essential nutrients from fresh organic residues as they decompose in soil. The active fraction of SOM has been most closely associated with nutrient supply. However, the stable, recalcitrant soil organic matter pool also improves soil fertility by holding plant nutrients and preventing them from leaching into the subsoil out of reach of plant roots. The SOM has a net negative charge and nutrients such as calcium, magnesium, potassium and ammonium (called cations) have a positive charge. The capacity of a soil to hold plant nutrients so that they are easily released or "exchanged" into the soil solution is measured by the cation exchange capacity (CEC). Soil CEC is estimated from the sum of exchangeable cations in the soil. Soil organic matter also binds plant micronutrients like iron, aluminum, zinc, copper and manganese by chelation—a chemical association of organic matter and micronutrients that keeps the micronutrients in a form available for plant uptake. Hence, SOM is the storehouse for major plant nutrients.

Organic matter has the ability to moderate major changes in the soil pH. Soil pH is a measure of acidity or alkalinity as determined by the amount of positively charged hydrogen (H^+) ions in the soil solution. Organic matter buffers the soil against major swings in pH by either taking up or releasing H^+ into the soil solution, making the concentration of soil solution H^+ more constant. The result is a stable pH close to neutral or suitable for the specific crop to be grown.

SOM binds the organic chemicals and pesticides, which affect their biological activity and toxicity. It also provides sites for microbes to colonize and decompose organic pollutants. Soil organic matter has been shown to protect plant roots from potentially harmful levels of aluminium, which occur in some acid soils and also from certain herbicides that are toxic to plants.

It improves the water percolation into and retention by soil. It involves the formation and maintenance of desirable soil structure. When fresh organic matter is added to the soil, soil microbes release long-chain sugars or polysaccharides relatively quickly. These polysaccharides promote formation of large or macro-aggregates. As the organic matter decomposes over the longer term, different sizes of aggregates are formed that are resistant to physical disruption. The number and diversity of stable soil aggregates are what give a soil an excellent physical structure. A soil with a good physical structure will have pores and channels of many sizes, from the tiniest channel to large spaces that will allow rainfall to penetrate without running off and taking soil with it. Soil organic matter helps to form and maintain the air passages and channels, protecting the soil from compaction. When a soil has plenty of spaces for air passage, holds water well without becoming too wet, is not compacted, and has a good physical structure for plant roots to grow and explore, refers good soil tilth. Organic matter is one of the key components to good tilth and more the SOM is the better the soil tilth.

The degree of surface soil aggregation will determine how tightly the soil particles are held during rain or wind storms. Stable soil aggregates resist movement by wind or water because they are larger than primary particles of silt or clay. Soil pores created by aggregation also promote water infiltration, thereby reducing runoff and the likelihood that soil particles will be transported with the water. Crop residues, vegetative cover or mulches on the soil surface will also slow down water as it impacts the soil during a rainstorm, giving it a better chance to seep into the soil. Preventing soil erosion conserves soil fertility because the most fertile soil is the top few inches.

SOM involves in the movement of tightly bound and normally insoluble metals with water soluble components of soil organic matter. It involves in the chelation of micronutrients and formation of spodic horizons, as it contributes to forming taxonomically different soils. It also absorbs solar radiation, which influences soil temperature. It provides colour to the soil and decreases its reflectance.

A soil rich in organic matter and regularly supplied with different kinds of soil organic matter will support a rich and varied population of soil organisms. Organic matter is the principle food source for secondary consumers. The most predominant functional group of secondary consumers are the decomposers: bacteria, fungi and actinomycetes.

There is an emerging awareness of global warming and its connection to human activities that promote greenhouse gas emissions (CO_2, CH_4 and N_2O). Carbon sequestration or storage in soil is considered a means to reduce greenhouse gas (CO_2 or CH_4) emissions.

Building and maintain higher soil organic matter levels should translate to lower levels of CO_2 being given off to the atmosphere. Research is underway to quantify the effects of different cropping systems and tillage practices on carbon sequestration.

Factors influencing the SOM in agroecosystem

In agroecosystem, factors that regulate organic matter functioning in soil are related to organic matter additions or input that influence particulate or macro organic matter and the relationships between organic matter and soil aggregates. The capacity of soil to store organic matter is related to the association of organic matter with clay and silt particles and soil microaggregates (Hassinik, 1997). Climate and soil types can significantly influence the accumulation and storage of soil organic matter due to the interaction of temperature and moisture on plant productivity and the ability of the soil mineral components to retain organic matter (Carter, 1996). Holding constant parent material and topography, the potential for organic matter accumulation increases with increasing precipitation and decreasing temperature (Cole *et al.,* 1993). Moist, warm or hot climates favour rapid soil organic matter decomposition relative to wet, cool climates.

The environmental and edaphic factors that control the activity of soil biota, and thus the balance between accumulation and decomposition of organic matter in the soil, are described below.

1. Temperature

Several field studies have shown that temperature is a key factor controlling the rate of decomposition of plant residues. Decomposition normally occurs more rapidly in the tropics than in temperate areas. Reaction rates doubled for each increase of 8-9°C in the mean annual air temperature. The relatively faster rate of decomposition induced by the continuous warmth in the tropics implies that high equilibrium levels of organic matter are difficult to achieve in tropical agro-ecosystems. Hence, large annual rates of organic inputs are needed to maintain an adequate labile soil organic matter pool in cultivated soils. Soils in cooler climates commonly have more organic matter because of slower mineralization (decomposition) rates (Ladd and Amato, 1985).

2. Soil moisture and water saturation

Soil organic matter levels commonly increase as mean annual precipitation increases. Conditions of elevated levels of soil moisture result in greater biomass production, which provides more residues, and thus more potential food for soil biota. Soil biological activity requires air and moisture. Optimal microbial activity occurs at near field capacity, which is equivalent to 60 % water-filled pore space (Linn and Doran, 1984). With the exception of the hyperhumid regions, the climates of vast areas of the humid, subhumid and semi-arid tropics are characterized by distinct wet and dry seasons. In the wet-dry tropics, large amounts of nitrate often occur in the surface soil during the first part of the rainy season (Greenland, 1958; Mueller-Harvey *et al.,* 1989). This accelerated nitrogen mineralization caused by a large increase in microbial activity is the result of the first few rains activating the labile soil organic matter.

3. *Soil texture*

Soil organic matter tends to increase as the clay content increases. This increase depends on two mechanisms. First, bonds between the surface of clay particles and organic matter retard the decomposition process. Second, soils with higher clay content increase the potential for aggregate formation. Macroaggregates physically protect organic matter molecules from further mineralization caused by microbial attack (Rice, 2002). For example, when earthworm casts and the large soil particles they contain are split by the joint action of several factors (climate, plant growth and other organisms), nutrients are released and made available to other components of soil microorganisms. Under similar climate conditions, the organic matter content in fine textured (clayey) soils is two to four times that of coarse textured (sandy) soils (Prasad and Power, 1997).

4. *Topography*

Organic matter accumulation is often favoured at the bottom of hills. There are two reasons for this accumulation: conditions are wetter than at mid- or upper-slope positions, and organic matter is transported to the lowest point in the landscape through runoff and erosion. Similarly, soil organic matter levels are higher on north-facing slopes (in the Northern Hemisphere) compared with south-facing slopes (and the other way around in the Southern Hemisphere) because temperatures are lower (Quideau, 2002).

5. *Salinity and acidity*

Salinity, toxicity and extremes in soil pH (acid or alkaline) result in poor biomass production and thus in reduced additions of organic matter to the soil. For example, pH affects humus formation in two ways: decomposition, and biomass production. In strongly acid or highly alkaline soils, the growing conditions for microorganisms are poor, resulting in low levels of biological oxidation of organic matter. Soil acidity also influences the availability of plant nutrients and thus regulates indirectly biomass production and the available food for soil biota. Fungi are less sensitive than bacteria to acid soil conditions.

6. *Vegetation and biomass production*

The rate of soil organic matter accumulation depends largely on the quantity and quality of organic matter input. Under tropical conditions, applications of readily degradable materials with low C:N ratios, such as green manure and leguminous cover crops, favour decomposition and a short-term increase in the labile nitrogen pool during the growing season. On the other hand, applications of plant materials with both large C:N ratios and lignin contents such as cereal straw and grasses generally favour nutrient immobilization, organic matter accumulation and humus formation, with increased potential for improved soil structure development. Plant constituents such as lignin and other polyphenols retard decomposition. In an experiment in southern Nigeria to compare management effects on soil organic matter accumulation, a three-year fallow with Guinea grass (*Panicum maximum*), which has a high lignin content, maintained a carbon level comparable to that under forest fallow. However, fallowing with leguminous species such as pigeon pea (*Cajanus cajan*) caused a significant decline in soil total C (Juo and Lal, 1977). However, in grassland ecosystems, up to two-thirds of organic matter is added through the decay of roots (Quideau, 2002).

Apart from these, placement of residues at depth (Carter *et al.,* 1998), quality of surface and soil incorporated crop residues (Juma, 1993) and soil friability (Watt and Dexter, 1998) also the other factors influence the accumulation of SOM.

SOM sustainability

Concern about the loss of organic matter from soils and the implications of this for the sustainable functioning of soils is not new, but the increasing demands that are being placed on our environment are leading to an urgent need to reassess the role played by soils in the development of sustainable patterns of land use. To sustain the SOM for long term, the key factors involved are:

1. The sequence and type of crop rotations have been widely shown to influence the SOM level. Rotation with annual crops reduce the SOM in greater level compared with perennial crops (Rees *et al.,* 2002).
2. Nutrient management of crops: Rate of mineralization is inversely proportional to SOM sustainability. While correlating the nutrient removal by crops, the loss should be equilibrated by residue management or by cover crops which will sustain the SOM.
3. Soil microbial diversity: Microorganisms in particular play an essential role in transformations of organic matter and nutrients that underpin many processes (Tiedje, 2000).

The sustainability of SOM has to date involved primarily chemical and physical treatment to improve structure, incorporate residues and stimulate decomposition. Thus management of SOM is very essential practice to sustain its level and activity in soil.

SOM management practices

1. Conservation agriculture

Conservation agriculture makes use of soil biological activity and cropping systems to reduce the excessive disturbance of soil and to maintain the crop residues on soil surface in order to minimize damage to the environment and provide organic matter and nutrients. It is based on four principles:

- minimal mechanical soil disturbance, mainly through direct seeding;
- permanent soil cover, organic matter supply through the preservation of crop residues and cover crops;
- crop rotation and efficient use of the soil profile;
- minimal soil compaction.

Although the principles are not new (except for that of minimal disturbance to the soil), it is the fact that they are applied together in conservation agriculture that generates positive outcomes. All the practices (minimal tillage, soil cover and crop rotation) are combined for synergy and added value. In the past, farmers may have tried but abandoned the use of cover crops or zero tillage because of weed problems or yield declines. There is also a need for improved weed control and rotations for biocontrol of pests and diseases and nutrient uptake. Integration of the conservation agriculture principles provides a favourable situation for both people and the environment, which has catalyzed successful expansion of the area under conservation agriculture worldwide.

2. Organic matter deposition

The reduction of soil disturbance through zero-tillage, the use of cover crops and the preservation of crop residues on the soil surface result in increased activity of the soil and in the accumulation of organic matter, mainly in the topsoil. An argument often on conservation agriculture is that it is only feasible in the humid and subhumid tropics and that the generation of sufficient biomass in semi-arid regions is the limiting factor to start implementing conservation agriculture. Mrabet (2000) reports higher yields through better water use and improved soil quality; the latter caused by an increase in soil organic C and N and a slight pH decline in the seedzone (Mrabet *et al.,* 2001a,b).

3. Increased carbon sequestration

World soils are important reservoirs of active C and play a major role in the global carbon cycle. As such, soil can be either a source or sink for atmospheric CO_2 depending on land use and the management of soil and vegetation. The conversion of native ecosystems (*e.g.* forests, grasslands and wetlands) to agricultural uses, and the continuous harvesting of plant materials, has led to significant losses of plant biomass and C (Davidson and Ackerman, 1993), thereby increasing the CO_2 level in the atmosphere. In particular, the practice of burning agricultural fields before cultivation has a disastrous effect on soil organic carbon content. The topsoil layer (0-5 cm) represented the greatest carbon loss (36 per cent) compared with the area that was not burned. Soil N stock in the same layer was reduced by 16 per cent. The carbon stock was reduced not only through burning, but because of the whole land-use management, especially a drastic reduction in diversity of species as monocropping was practiced.

Organic manures

Organic manures are source of SOM, which are depleted due to continuous cropping. They not only increase the SOM content, but also provide nutrients to the existing crops. There is no end to the number of possible organic amendments available for soil. Many are currently considered "waste" products or by-products from farm enterprises, food processors, municipalities or industry. Bulky organic manures are rich sources of organic matter. These manures include green manures, farmyard manure, farm compost, urban compost, night-soil, sludge and other bulky sources of organic matter. Out of these, green manure, farmyard manure and composts are by far the most important and very widely used. Crop residues also make significant contribution to organic matter content of soil. They can be applied raw or following some kind of processing like anaerobic digestion, composting or drying/pelletizing.

Sources of organic manures

The FAO soil map of the world estimate the presence of 22 x 10^{14} kg organic carbon in global soils made up from 18 x 10^{14} kg carbon in mineral soil and 4 x 10^{14} kg carbon in the surface matter of peat soil (Bohn, 1982). The annual return of plant and animal residues due to the soil varies greatly with climatic region and the type of vegetation or land use. About 919 million tonnes of cereals was produced in the developed and developing countries indicating that cereal residue production probably exceeded 1000 million tonnes per annum. Common organic amendments (Table 2) availabl nclude:

1. On-farm wastes: manure, crop residues, spoiled straw, hay and silage.
2. Municipal wastes: yard debris and biosolids (sewage sludge).
3. Organic wastes from food processors: vegetable, meat, fish, dairy.
4. Organic wastes from paper mills and the timber industry.
5. Composts.

Table 2: Common organic amendments and their C:N ratios

S. No.	Organic amendments	C:N ratio
High in carbon		
1.	Corn stalks	60-70
2.	Straw	40-150
3.	Corn silage	40
4.	Fall leaves	30-80
5.	Sawdust	200-700
6.	Brush, wood chips	100-500
7.	Bark (paper mill waste)	100-130
8.	Newspaper	400-800
9.	Cardboard	500
10.	Mixed paper	150-200
High in Nitrogen		
11.	Hay	15-30
12.	Dairy manure	5-25
13.	Poultry manure	5-15
14.	Hog manure	10-20
15.	Cull potatoes	18
16.	Vegetable wastes	10-20
17.	Coffee grounds	20
18.	Grass clippings	15-25
19.	Municipal biosolids	9-25

Animal manures and biosolids (sewage sludge) are good examples of organic amendments with fertilizer value. Both can supply N, P and K needs of many crops because greater than 25% of their total N, P and K contents are in forms readily available for crop uptake (Table 3). Amendments like municipal yard waste, bark and composts are examples of soil conditioners. They are not considered fertilizer substitutes, but mainly they improve soil properties by building soil organic matter.

All organic matter contains some carbon and some nitrogen. The relative amount of each present (the carbon to nitrogen ratio) will determine whether nitrogen, and sometimes other nutrients, will be tied up in the decomposition process. Decomposition of materials with a low C:N ratio like animal manures (less than about 20:1) will release nitrogen, while the decomposition of high C:N materials such as straw or sawdust, will require that soil microbes use the amendment N for their own metabolic needs.

Table 3: Nutrient status of some agro industrial wastes

Source	N (%)	P (%)	K (%)
Coir pith	0.26	0.05	0.84
Press mud	1.15	2.40	1.98
Sea weed residue	1.50	0.50	4.20
Cotton mill waste	1.40	0.60	1.20
Saw dust	0.25	0.20	1.00
Sewage sludge	1.50	2.00	1.50
Bagasse	0.25	0.12	0.40
Biogas slurry	1.60	1.50	1.20
Mushroom spent wash	1.68	0.60	0.63

The common practices to enhance the SOM level through organic manures are described below:

1. Crop residue management

In systems where crop residues are managed well, they:

1. add soil organic matter, which improves the quality of the seed bed and increases the water infiltration and retention capacity of the soil, buffers the pH and facilitates the availability of nutrients;
2. sequester (store) C in the soil;
3. provide nutrients for soil biological activity and plant uptake;
4. capture the rainfall on the surface and thus increase infiltration and the soil moisture content;
5. provide a cover to protect the soil from being eroded;
6. reduce evaporation and avoid desiccation from the soil surface.

An even distribution of residues: (i) provides homogenous temperature and humidity conditions at sowing time; (ii) facilitates even sowing, germination and emergence; (iii) minimizes the development of pests and diseases; and (iv) reduces the emergence of weeds through allelopathic effects.

The most appropriate method for managing crop residues depends on the purpose of the crop residues and the experience and equipment available to the farmer. Where the aim is to maintain a mulch over the soil for as long as possible, the biomass is best managed using a knife roller, chain or sledge in order to break it down but not kill it. Where the decomposition process should commence immediately in order to release nutrients, the residues should be slashed or mown and some N applied because dry residues have a high C:N ratio. However, in order to avoid nitrate emission, urea should not be broadcast on the surface but injected where possible.

In India, large numbers of crops are growing and after harvesting economic parts, the remaining portion is mostly wasted except for a few crops. The potential crop residues of major cereals, pulses, oilseeds and commercial crops for recycling of valuable plant nutrients for sustained crop production is enormous. Average nutrients of crop residues, presented in Table 4 showed high potential for soil organic matter addition.

Table 4: Estimate of available and reliable nutrients from crop residues

Crop	Residue ('000 t)	Nutrient content N (%)	P (%)	K (%)	Nutrient potential ('000 t)
Rice	80744	0.61	0.09	1.15	1493.8
Wheat	44987	0.48	0.07	0.98	688.3
Sorghum	11563	0.52	0.10	1.21	216.2
Maize	6219	0.58	0.09	1.25	199.4
Pearl millet	8283	0.45	0.07	0.95	121.6
Barley	3180	0.52	0.08	1.25	88.8
Sugarcane	15645	0.45	0.08	1.20	270.7
Potato	5062	0.52	0.09	0.85	73.9
Groundnut	9580	1.65	1.20	1.23	277.3

2. Applying animal manure or other carbon-rich wastes

Any application of animal manure, slurry or other carbon-rich wastes, such as coffee-berry pulp, improves the organic matter content of the soil. In some cases, it is better to allow a period of decomposition before application to the field. Any addition of carbon-rich compounds immobilizes available N in the soil temporarily, as microorganisms need both C and N for their growth and development. Animal manure is usually rich in N, so N immobilization is minimal. Where straw makes up part of the manure, a decomposition period avoids N immobilization in the field.

3. Compost

Composting is a technology for recycling organic materials in order to achieve enhanced agricultural production. Biological and chemical processes accelerate the rate of decomposition and transform organic materials into a more stable humus form for application to the soil. Composting proceeds under controlled conditions in compost heaps and pits. Composting can complement certain crop rotations and agroforestry systems. It can be used efficiently in planting pits and nurseries. It is very similar in composition to soil organic matter. It breaks down slowly in the soil and is very good at improving the physical condition of the soil (whereas manure and sludge may break down fairly quickly, releasing a flush of nutrients for plant growth). In many circumstances, it takes time to rejuvenate a poor soil using these practices because the amount of organic material being added is small relative to the mineral proportion of the soil. Table 5 showed the advantages of composted manures over raw manures (Coyne, 1999).

Table 5: Qualities of raw and composted manure

Raw manure	Composted manure
High available nutrients: N, P, K	Low available nutrients, esp. N
Heterogeneous and high volume	Homogeneous and low volume
Very active biologically	Biologically stable
Strong odour	Non-offensive odour
Wet	Moist-dry
May contain weed seeds and pathogens	Pathogens and weed seeds killed

The obvious and traditional use of compost is a known soil amendment which can benefit soil by supplying plant nutrients and also improving the traditional method of composting by hastening process through inoculation of celluloytic fungi and enrichment of compost from rural and urban wastes which would act as a good source of plant nutrients. The nutrient content of composts from qualitative stand point is very favorable and being biological in origin, it contains all essential plant nutrients. Normally composts prepared from farm wastes are superior in organic matter and the major nutrients N, P, K than those prepared from urban wastes (Mishra and Hesse, 1982). The average nutrient status of some common organic manures and composts from agricultural residues are presented in Table 6.

Table 6: Average nutrient status of organic manures

Compost	N (%)	P (%)	K (%)
Farm yard manure	0.80	0.41	0.74
Poultry manure	2.87	2.93	2.35
Rural compost	1.22	1.08	1.47
Urban compost	1.24	1.92	1.07
Mixed farm waste compost	0.87	0.59	2.22
Mixed dry residue compost	0.90	0.45	1.95
Cotton stalk compost	1.61	0.48	3.38
Water hyacinth compost	2.00	1.00	2.30
Paddy straw compost	1.59	1.34	3.37
Dry grass / weeds compost	1.90	0.55	1.09
Cotton waste / Groundnut husk	1.62	1.04	1.26
Lantana stalk and leaf compost	1.55	0.52	1.07
Wheat straw compost	2.90	2.05	0.90
Maize stalk compost	1.99	1.30	1.01
Sugarcane trash compost	2.73	1.81	1.31
Vermicompost	1.60	2.20	0.67

4. Mulch or permanent soil cover

One way to improve the condition of the soil is to mulch the area requiring amelioration. Mulches are materials placed on the soil surface to protect it against raindrop impact and erosion and to enhance its fertility (FAO, 1995). Crop residue mulching is a system of maintaining a protective cover of vegetative residues such as straw, maize stalks, palm fronds and stubble on the soil surface. The system is particularly valuable where a satisfactory plant cover cannot be established rapidly when erosion risk is greatest. Mulching adds organic matter to the soil, reduces weed growth and virtually eliminates erosion during the period when the ground is covered with mulch. Mulch affects the soil life. Holland and Coleman (1987) have demonstrated that litter placement on the soil surface increased the ratio of fungi to bacteria—the reason being that fungi have a higher carbon assimilation efficiency than bacteria. In addition, it encourages mixing effects of macrofauna that pull the materials into surface layers of the soil.

Microbial inoculants

SOM and soil microbial activity

One of the important functions of the soil biological community is managing nutrients. Soil organisms continually transform nutrients among many organic and inorganic forms. Plants primarily need simple inorganic forms of each nutrient. Soil organisms create many of these plant-available nutrients and help to store nutrients in the soil as organic compounds. The three important functions of soil microorganisms with respect to nutrient cycle are decomposition, mineralization and immobilization. Decomposition is the breakdown of plant and animal residue into different organic and inorganic compounds. Soil organisms decompose organic matter more quickly under warm, moist conditions than under cold or dry conditions. As part of the decomposition process, many bacteria and fungi produce humic acids. In the soil, these acids chemically combine with each other to form large molecules of stabilized organic matter. This formation of large molecules is both a biological and chemical process. When soil organisms convert organic matter into inorganic, plant-available nutrients, they are said to be mineralizing nutrients. Protozoa and nematodes mineralize and excrete several kilos of ammonium (NH_4^+) per acre per day. Most is snatched up by other soil organisms, but some is used by plants. The reverse of mineralization is immobilization—the conversion of inorganic compounds into organic compounds. Soil organisms consume inorganic molecules and incorporate them into their cells. Because immobilized nutrients are parts of soil organisms, they do not move easily through the soil and are unavailable to plants. Bacteria and fungi are responsible for large amounts of immobilization. To facilitate and rapid decomposition, the soil microbial community composition is very important.

Like the above-ground ecosystem, the soil community is not just a collection of individual species, but a complex, interacting food web. Decomposition of a single compound may require several organisms. The creation of aggregates involves a mix of physical and chemical processes and the activity of many types of organisms. As the complexity of the food web increases, productivity of the soil tends to increase. It is not clear how much complexity is needed, but there are several reasons why complexity is thought to be beneficial. Other benefits of complexity may include improved nutrient cycling, decomposition, and disease control. When many different kinds of organisms are present, many organic compounds and potential pollutants can be degraded, and many competitors and predators are present to control pest populations.

Microbial inoculants for biomanures

The microorganisms needed for composting are found throughout the natural environment. They are present in compost feedstock as well as in the water, air, soil and machinery the feedstock and compost are exposed to during processing. These sources ensure a high diversity of microorganisms, which helps to maintain an active microbial population during the dynamic chemical and physical processes of composting, such as shifts in pH, temperature, water, organic matter and nutrient availability. Only on rare occasions will the addition of microorganisms be warranted. The majority of microorganisms responsible for the formation of compost are aerobes in that they require or work best in the presence of oxygen.

Many difficulties associated with composting may be traced to insufficient oxygen levels to support the decomposition of compost feedstock. Compost microbes also require a moist environment because they live in the water films surrounding composting organic matter particles. A 50 to 60 per cent moisture content is optimal. Fungi and actinomycetes penetrate throughout the composting material, decomposing both chemically and mechanically the more recalcitrant organic matter fraction such as lignins and cellulose. Fungal hyphae physically stabilize the compost into small aggregates, providing the compost with improved aeration and drainage. Fungi number between 0.01 and 1 million propagules per gram of soil. About 70,000 different species of fungi have been described worldwide, but an estimated 1 million additional species remain undiscovered and undescribed. Ecologically, fungi play a vital role in breakdown of dead plant materials. The most numerous biological component of compost is the bacteria. Although they often can exceed 1 billion microorganisms per gram of soil, bacteria do not contribute as much to the overall microbiological mass as fungi because of their relatively small size. Although bacteria exist as individuals and do not form filaments, they also contribute to the stabilization of aggregates through the excretion of organic compounds that bind adjacent organic matter and soil particles together. Bacteria are typically associated with the consumption of easily degraded organic matter. They are the dominant population throughout the entire composting process, whereas the actinomycetes and fungi typically proliferate in the later stages.

The crop residues and other cellulose rich organic wastes can be inoculated with cellulolytic and lignolytic microorganisms either *in situ* or in composting yard to hasten the decomposition so as to reduce the nutrient immobilization while applying to soil. Yadav *et al.* (1982) used composite cellulolytic fungi consisting *Pacelomyces fusisporus, P. varioti, Penicillium chrysogenum, Sclerotium rolfsii, Aspergillus fumigatus, Humicola lanuginosa* and *Mucor mieehii,* decomposed the cellulosic waste rapidly when compared the unifungal inoculants. The mixed inoculant reduced the C:N ratio and increased the N content of the manure in short incubation period. Guar *et al.* (1982) used different fungal inoculants and among them, *Aspergillus* and *Penicillium* inoculation hasten the decomposition. Balachandar and Kannaiyan (1999) used seven cellulolytic bacterial composite inoculant to prepare the biomanure from cellulose rich viscose factory waste. During these inoculation studies, the inoculants used as starter microflora of decomposition, later the native flora from the waste material built-up and increased the decomposition rate.

The available nitrogen, phosphorus and potassium content of cellulosic organic manures, when incubated with cellulolytic microbial inoculants have increases more than two folds than the original nutrient status, because of more microbial activity, enzyme production and fast narrow down of C:N ratio, comparing the uninoculated biomanures (Sarojini and Mathur, 1994; Balachandar and Kannaiyan, 1996). Similarly the humus content of the organic manures after final product was also increased by microbial inoculation and incubation (Gaur, 1987).

Inoculation of suitable cellulolytic and lignolytic strains has been reported to hasten the rate of decomposition, which in turn leads to the enrichment of nutrients in final product (Tiwari *et al.,* 1989). Lignin is the most recalcitrant material present in organic wastes and decomposes only at the later stage of the decomposition (Manna *et al.,* 2003). *Phanerochaeta chrysosporium*, which is considered as the most effective lignin-decomposing white rot fungi (Chahal, 1994), is supposed to enhance the rate of decomposition. *Trichoderma viridae,* producing cellulase enzyme and having bioaccumulation ability is also considered as consistently effect the inoculants for composting (Bhardwaj and Guar, 1985). Kaushik and Garg (2004) reported that inoculation of N-fixing bacteria during composting increased the nitrogen content of stabilized product. Beauchamp *et al.* (2006) reported that nitrogen fixing ability of bacteria depended on the organic wastes used. Pramanik *et al.* (2007) used mixed inoculant containing *Bacillus polymyxa, Trichoderma viridae* and *Phenerochaeta chrysosporium* for hastening the municipal solid waste by earth worms. When inoculated with this mixed inoculant, the quality of vermicompost prepared from the organic wastes have showed improved nutrient status, enzyme activity and microbial activity than uninoculated vermicomposts.

Nutrient amendments for biomanures

Due to high cost of fertilizers, there is a growing awareness towards exploitation and utilization of available organic resources in the country. Organic manures are known to improve physical, physiochemical and biological properties of soil but because of their low nutrient content and slow acting nature, organic manures alone may not be able to meet the high nutritional requirements of crops and therefore fertilizers have to be applied. It is generally accepted that some of the plant nutrients when added to soils in organic form, their efficiency is low as compared to their application along with organic manures or inorganic forms which have higher manurial value. The recycling of renewable organic wastes to meet the challenges of 21st century for fertilizer, food and feed is utmost economic importance by improving the traditional method of composting by hastening the process through inoculation of cellulolytic fungi and enrichment of compost both from rural and urban wastes which is good source of plant nutrients.

The cellulosic wastes cannot be applied as such because of their wider C:N ratio which limit the decomposition process. They interfere with availability of essential nutrients required by the crop through immobilization into organic forms. On the other hand, the enrichment of composting materials with some amendments enhanced the rate of decomposition and improved the nutrient status of compost (Enweizer, 1976). Addition of cheep nitrogen and phosphorus sources could improve the manures by narrowing down the C:N ratio and enhanced nutrients content (Dash *et al.,* 1981). Addition of cow dung is a good practice of manure preparation, as it provides wide diversified microbial sources for decomposition (Singh, 1987). Kapoor *et al.* (1983) found addition of rock phosphate along with farm wastes reduced the total weight of the manure by 50%. Singh (1987) reported that enrichment of farm wastes with cow dung, rock phosphate and molasses hasten the decomposition and recorded the weight loss of 40% after a month incubation. Kapoor *et al.* (1983) used free-living nitrogen fixing bacterium, *Azotobacter chroococcum* and phosphate solubilizing organisms, *Aspergillus awamori* and *Bacillus polymyxa* for nutrient enrichment along with rock phosphate and found that enrichment of P solubilizers and *Azotobacter* reduced the C:N ratio rapidly and increased the nitrogen and humus content of the manure.

Swaminathan *et al.* (1982) conducted experiments with urea and single super phosphate enrichment of manure prepared from farm yard manure and their influence on growth and nutrient uptake of maize (Table 7). Among the different manures applied, urea and SSP enriched manure application recorded maximum plant growth and N and P uptake than other fertilization in maize crop.

Table 7: Influence of nutrient enrichment (N and P) of composts on plant growth and nutrient uptake of maize

Source of nutrients	Shoot weight (g/pot)	Root weight (g/pot)	N uptake (mg/pot)	P-$_2$O$_5$ uptake (mg/pot)
No enrichment of manure (pit)	3.73	0.85	94.43	19.82
No enrichment of manure (heap)	4.32	1.04	110.62	21.62
Urea + SSP enriched manure (pit)	4.71	1.17	127.75	25.35
Urea + SSP enriched manure (heap)	5.45	1.33	150.20	24.55
Urea + SSP fertilizers	5.73	1.37	160.44	31.57
CD (5%)	0.09	0.02	2.49	1.25

Similarly, composting low grade rock phosphate with agricultural waste is known to increase the solubility of insoluble phosphates (May *et al.,* 1986). Although the extent of solubilization can vary with the kind of waste, rate of organic matter decomposition during composting are of primary importance. The production of humic substance and their retention capacities for phosphorus and calcium released during composting of wheat straw with low grade rock phosphates revealed that humic acid production increased with increased incubation time, while fulvic acid decreased after 30 days (Singh and Amberger, 1990).

Nutrients and crop productivity

Addition of organic manures to soil enhanced the microbial activity particularly the heterotrophs. The organic manures greatly influence the free-living nitrogen-fixing bacterial population, particularly, *Azotobacter* (Table 8). The release of nutrients from decomposing organic matter led to the source for heterotrophic bacterial community which in turn increased several folds and the microbial biomass carbon content will be increased. Conversion of organic carbon to microbial carbon is a good sign for plant nutrients availability as the microbial communities conserve the nutrient flow.

Table 8: Influence of different doses of organic manures on soil *Azotobacter* population (Ramaswami *et al.*, 1979)

Treatment	Azotobacter population (cfu x 10^4 per g soil)
Control	2.0
NPK (180-90-90) (kg/ha)	18.7
Straw (10 t/ ha)	35.3
Straw (15 t/ ha)	152.0
Straw (10 t/ ha) + 90 kg/ha rock phosphate	69.0
Straw (15 t/ ha) + 90 kg/ha rock phosphate	352.0

Similarly the organic manure application to soil mostly suppressed the pathogenic fungal community as the organic manures are good source of decomposing fungal community and antagonists.

The other beneficial effect of organic manure application is continuous supply of major nutrients like N and P. In organic manure applied soils, mineralization of applied organic N was 2 to 11 per cent under anaerobic and aerobic soil conditions. The rate of N mineralization from organic manures mainly depends on the C:N ratio of the manure and the microbial diversity of the soil. N management in problem soils like acidic and sodic soils is very difficult which can be enhanced by organic manures. The organic manure application reduced the rate of mineralization in acidic soils and thereby reduced the N – loss (Gupta and Bajpai, 1978).

Organic manure application enhanced the P use efficiency and P solubility as it has high pH buffering activity. Supply of organic P, which is less influenced by pH variations and maintaining neutral pH by organic manures led to the increased HPO_4 and H_2PO_4 levels in soil solution are the major reasons for enhanced P availability by organic manures. Similarly production of high amount of organic acids during decomposition of organic manures also led to solubilize the insoluble P to soluble forms (Pareek and Guar, 1983). Tripathi and Minhas (1991) inferred that addition of FYM enhanced the solubility of all the phosphate fractions probably by interfering in some way with the formation of less soluble octacalcium from dicalcium phosphate. Phosphorus might have been released through the acidifying effect of CO_2 and the action of organic chelates produced by microbial decomposition.

The continuous recycling of crop residues for seven years in rice-wheat rotation markedly enhanced the soil properties has been reported by Bhat *et al.* (1991). The organic carbon status was significantly increased when crop residues were incorporated. The dry matter production of wheat crop was increased in which crop residues had been incorporated for seven years than in soils where the crop residues are burnt or removed. The response of crops to organic manures depending on the soil quality varied from less to greater extent. Maximum of 50-60% more yield was recorded in various crops comparing to the fertilizer applied crops. Similarly the response of wheat crop to 120 kg N/ha yielded almost the same of the wheat crop receiving compost with 60 kg N/ha.

Conclusion

The soil organic matter exerts a profound influence on the physical, chemical and biological properties of the soil. In encourages soil granulation and improves the water holding capacity and drainage properties of soil. On mineralization, the organic matter releases plant nutrients slowly which are less subject to leaching losses. The humus content of the soil adsorbs nutrients and make them available to higher plants as and when needed, thus functioning a buffer of nutrients. The organic matter can also block the fixation of phosphorus and can be neutralized to some extent, element such as aluminium considered poisonous to plant growth. Similarly SOM adsorbed some of the toxic chemicals like herbicides and thus protect from phytotoxicity to crop plants and to some extent involved the carbon sequestering by humus formation. Thus the SOM is the most important component of the soil which decides the productivity of the soil.

The major source of organic carbon entering to the soil is the plant residues and composts. Complex materials added to the soil undergo physical, chemical and biological changes during decomposition. Maintaining the soil organic matter is very critical, which can be done by crop residue management, animal and other industrial organic manures, mulch management, composting and cover cropping. The biomanure prepared from composting different organic waste both industrial and agricultural wastes serve as high nutrient organic fertilizer in modern organic agriculture. To reduce the composting time and to increase the quality of the compost, many cellulolytic bacteria and fungi are used as inoculants. There are several nutrient amendments used includes, rock phosphate, SSP animal waste, poultry waste, goat manure urea, *Azotobacter*, phosphobacteria, cowdung to enhance the decomposition and to increase the nutrient values of organic manures.

Maintaining optimum dose of organic matter in soil through biomanure facilitates the nutrient supply to crops, maintaining soil pH, maintaining soil microbial diversity, reduces the disease incidences, increased soil health and sustainable crop productivity. The addition of organic materials sources is a key factor in maintaining soil fertility and long term sustainability for higher as well as efficient crop productivity.

Future thrust

- Processing the organic wastes for rapid decomposition and nutrient rich biomanures for various cropping systems,
- Developing composite inoculants with organic waste specificity for hastening the decomposition,

- Development of *in situ* microbial carbon sequestering in soil through efficient microbial communities,
- Studying the functional microbial community, which build and maintain the soil organic matter,
- Development of integrated approach to hasten the waste decomposition through physical, chemical and biological methods,
- Development of biomanures with potential N_2-fixing and P-solubilizing organisms to enhance the nutrient supply and hasten the decomposition,
- Development of *in situ* composting microorganisms for residue management,
- Development of packages for residue management for every cropping system to maintain the soil organic matter,

Are some of the important concepts and future thrusts needed for maintaining the soil organic matter level and to supply organic manures to sustain the soil fertility.

References

Balachandar, D. and Kannaiyan, S. 1996. Effect of inoculation of cellulolytic bacteria and enrichment of amendments on decomposition of viscose factory wastes. Paper presented in *National Seminar on Environmental Biotechnology* held at Bharathidhasan Univ., Trichy, pp. 100.

Balachandar, D. and Kannaiyan, S. 1999. Microbial decomposition of cellulosic waste and their biomanure effect on rice. In: *Bioresources Technology for Sustainable Agriculture* (ed). Kannaiyan, S., Associated Publishing Co., New Delhi, pp. 147-161.

Beauchamp, C.J., Levesque, G., Prevost, D. and Chalifour, F.P. 2006. Isolation of free-living dinitrogen-fixing bacteria and their activity in compost containing de-inking paper sludge. *Bioresource Technol.*, **97**: 1002-1011.

Bhardwaj, K.K.R. and A.C. Gaur. 1985. *Recycling of Organic Wastes*. Indian Council of Agricultural Research, New Delhi, pp. 54-58.

Bhat, A.K., V. Beri and B.S. Siddhu. 1991. Effect of long term recycling of crop residues on soil productivity. *J. Indian Soc. Soil Sci.*, **39**: 380-382.

Bohn, H. 1982. Estimate of organic carbon in world soils II. *Soil Sci. Soc. Amer. J.*, **11**: 18.

Carter, M.R. 1996. Analysis of soil organic matter storage in agroecosystem. In: *Structure and organic matter storage in agricultural soils*. (eds.) Carter, M.R. and D.A. Stewart, CRC Press, Florida, pp. 3-11.

Carter, M.R., E.G. Gregorich, D.W. Anderson, J.W. Doran, H.H. Janzen and F.J. Pierce. 1998. Concept of soil quality and their significance. In: *Soil Quality for Crop Production and Ecosystem health* (eds.) Gregorich, E.G. and M.R. Carter, Elsiever Publ., The Netherlands, pp. 1-17.

Chahal, D.S. 1994. Biological disposal of lignocellulosic wastes and alleviation of their toxic effluents. In: *Biological Degradation and Bioremediation of Toxic Chemicals*. (ed.) Chaudhary, G.R. Chapman and Hall, London, pp. 364-365.

Cole, V.C., K. Paustian, E.T. Elliott, A.K. Metherell, D.S. Ojima and W.J. Parton. 1993. Analysis of agroecosystem carbon pool. *Water Air Soil Pollut.*, **70**: 357-371.

Copperband, L. 2002. *Building soil organic manure with organic amendments*. Centre for Integrated Agrl. Systems, Madison, pp. 16.

Coulding, K.W.T., D.V. Murphy, A. Macdounald, E.A. Stockdale, J.L. Gaunt, L. Blake, G. Ayaga and P. Brookes. 2001. Role of soil organic matter and manures in sustainable nutrient cycling. In: *Sustainable Management of Soil Organic Matter*. (eds.) Rees, R.M., C.D. Campbell and C.A. Watson, CABI publ., Oxon, UK, pp. 221-232.

Coyne, M. 1999. *Soil Microbiology: An Experimental Approach*. Delmar Publ., New York, pp. 462.

Dash, R.N., S.K. Mohantly and S. Patnaik. 1981. Chemical composition and reactivity of some Indian rock phosphates. *J. Indian Soc. Soil Sci.*, **29**: 559-562.

Davidson, E.A. and I.L. Ackerman. 1993. Changes in soil carbon inventories following cultivation of previously untilled soil. *Biogeochem.*, **20**: 161-193.

Enweizer, W.O. 1976. The mineralization of N and P in organic materials of varying C:N ratio and C:P ratio. *Plant Soil*, **44**: 237-240.

FAO. 1995. *Tillage systems in the tropics. Management options and sustainability implications.* FAO Soils Bulletin No. 71. Rome.

Greenland, D.J. 1958. Nitrate fluctuations in tropical soils. *J. Agric. Sci.*, **50**: 82-92.

Guar, A.C. 1987. Recycling of organic waste by improved techniques of composting and other methods. *Resources Conservation*, **13**: 157-174.

Guar, A.C., K.V. Sadhasivam, S.P. Magu and R.S. Mathur. 1982. Role of mesophilic fungi in composting. *Agric. Wastes*, **4**: 45-53.

Gupta, B.R. and P.D. Bajpai. 1978. Organic matter in soil. *J. Indian Soc. Soil Sci.*, **26**: 173.

Hassinik, J. 1997. The capacity of soil to preserve organic carbon and N by their association with clay and silt particles. *Plant Soil*, **191**: 77-87.

Holland, E.A. and D.C. Coleman. 1987. Litter placement effects on microbial and organic matter dynamics in an agro-ecosystem. *Ecol.*, **68**: 425-433.

Juma, N.J. 1993. Interrelationships between soil structure/texture, soil biota/soil organic matter and crop production. *Geoderma*, **57**: 3-30.

Juo, A.S.R. and R. Lal. 1977. The effect of fallow and continuous cultivation on physical and chemical properties of an Alfisol in western Nigeria. *Plant Soil*, **47**: 507-584.

Kapoor, K.K., K.S. Yadav, D.P. Singh, M.M.Mishra and P. Tauro. 1983. Enrichment of compost by *Azotobacter* and phosphate solubilizing microorganisms. *Agric. Wastes*, **4**: 453-460.

Kaushik, P. and V.K. Garg. 2004. Dynamics of biological and chemical paramerers during vermicomposting of solid textiler mill sludges mixed with cow dung and agricultural residues. *Bioresource Technol.*, **4**: 203-209.

Ladd, J.N. and M. Amato. 1985. Nitrogen cycling in legume cereal rotation. In: *Nitrogen Management in Farming System in Humid and Sub-humid Tropics.* (eds.) Kang, B.T. and J. Van der Heide, Haren, The Netherlands, pp. 105-127.

Linn, D.M. and J.W. Doran. 1984. Effect of water-filled pore space carbon dioxide and nitrous oxide production in tilled and nontilled soils. *Soil Sci. Soc. Am. J.*, **48**: 1267-1272.

Manna, M.C., S. Jha, P.K. Ghosh and C.L. Acharya. 2003. Comparative efficacy of three epigeic earthworms under different deciduous forest litters decomposition. *Bioresource Technol.*, **88**: 197-206.

Martin, J.P. and K. Haiden. 1971. Microbial activity in relation to soil humus formation. *Soil Sci.*, **111**: 54-63.

May, P., D. Sayay and L. Andre. 1986. Chemical and microbiological solubilization of rock phosphates. *Comtes/Redus Seances de L Academie d'Agriculture de France,* **72**: 81-90.

Mishra, R.V. and P.R. Heese. 1982. *Comparative analysis of organic manures.* Project Field Document No.24. FAO/UNDP, Rome, Italy.

Monreal, C.M., H.R. Schulten and H. Kodama. 1997. Age, turn over and molecular diversity of soil organic matter in aggregates of a Gleysol. *Canadian J. Soil Sci.,* **77**: 379-388.

Mrabet, R. 2000. Long-term no tillage influence on soil quality and wheat production in semiarid Morocco. Paper presented at the 15th ISTRO Conference, USA, 2-7 July 2000.

Mrabet, R., K. Ibno-Namr, F. Bessam, and N. Saber. 2001a. Soil chemical quality and implications for fertilizer management after 11 years of no-tillage wheat production systems in semiarid Morocco. *Land Deg. Dev.,* **12**: 505-517.

Mrabet, R., N. Saber, A. El Brahli, S. Lahlou and F. Bessam. 2001b. Total, particulate organic matter and structural stability of a Calcixeroll soil under different wheat rotation and tillage systems in semiarid areas of Morocco. *Soil Till. Res.,* **57**: 225-235.

Mueller-Harvey, I., A.S.R. Juo and A. Wild. 1989. Mineralization of nutrients after forest clearance and their uptake during cropping. In: *Mineral nutrients in tropical forest and savannah ecosystems.* (ed.) Proctor, J., Blackwell Scientific Publications, London, UK. pp. 315-324.

Pareek, R.P. and A.C. Guar. 1983. Association of certain fungi with paddy seeds and the effects of seed dressing chemicals on them. *Curr. Sci.,* **42**: 278.

Paul, E.A. and F.E. Clark. 1996. *Soil microbiology and biochemistry.* Academic Press, Sandiego, CA.

Pramanik, P., G.K. Ghosh, P.K. Ghosal, P. Banik. 2007. Changes in organic – C, N, P and K and enzyme activities in vermicompost of biodegradable organic wastes under liming and microbial inoculants. *Bioresource technol.,* **98**: 2485–2494.

Prasad, R. and J.F. Power. 1997. *Soil fertility management for sustainable agriculture.* Lewis Publishers, New York, USA, pp. 356.

Quideau, S.A. 2002. Organic matter accumulation. *In: Encyclopedia of soil science,* Marcel Dekker Inc, New York, USA, pp. 891-894.

Ramaswami, P.P., K. Andi and K.K. Krishnamoorthy. 1979. Utilization of rice straw for enhancing *Azotobacter* population in rice soil. *Farm. Agric.,* **11**: 47-49.

Rees, R.M., B.C. Ball, C.D. Campbell and C.A. Watson. 2001. Sustaining soil organic matter. In: *Sustainable Management of soil organic matter.* (eds.) Rees, R.M., C.D. Campbell and C.A. Watson, CABI Publ., Oxon, UK, pp. 413-425.

Rice, C.W. 2002. Organic matter and nutrient dynamics. *In: Encyclopedia of soil science,* Marcel Dekker Inc, New York, USA, pp. 925-928.

Sarojini, V. and R.S. Mathur. 1994. Biodegradation of paddy straw inoculated with cellulolytic fungi from palm oil mill effluent. *World J. Microbiol. Biotechnol.,* **8**: 614-617.

Singh, C.P. 1987. Preparation of high grade compost by an enrichment technique. 1. Effect of enrichment on organic matter decomposition. *Biol. Agric. Hort.,* **5**: 41-47.

Singh, C.P. and A. Amberger. 1990. Humic substances in straw compost with rock phosphate. *Biol. Wastes,* **31**: 165-174.

Swaminathan, K.R., P. Rajasekaran, R. Manian and S. Kamaraj. 1982. Enrichment of nutrients in organic wastes through anaerobic digestion. *Madras Agric. J.,* **69**: 590-592.

Tiedje, J.M. 2000. 20 years since Dunedin: the past and future of microbial ecology. In: *Microbial biosystems: New Frontiers*, (eds.) Bell, C.R., M. Brylinski and P. Johnson-Green, Halifax, Canada, pp. 933-939.

Tiwari, S.C., B.K. Tiwari and R.R. Mishra. 1989. Microbial populations, enzyme activities and nitrogen, phosphorus and potassium enrichment in earthworm casts and in surrounding soil of a pineapple plantation. *Biol. Fertil. Soils*, **8**: 178-182.

Tripathi, D. and R.S. Sinhas. 1991. Influence of fertilizer phosphorus and farm yard manure on transformation of inorganic phosphates. *J. Indian Soc. Soil Sci.,* **39**: 472-476.

Watts, C.W. and A.R. Dexter. 1998. Soil friability: Theory, measurement and the effects of management and organic carbon content. *Eur. J. Soil Sci.*, **49**: 73-84.

Yadav, K.C., M.M. Mishra and K.K. Kapoor. 1982. The effect of fungal inoculation on composting. *Agric. Wastes*, **4**: 329-333.

Soil Microbiology and Soil Biotechnology : Vol. 1
Editor : S. Kannaiyan
Associated Publishing Co., New Delhi, 2010, pp. 31-42

3

Microbial Decomposition of Pine Needle Litter

G. Selvakumar[1] *and S. Kundu*[2]

Introduction

Pine (*Pinus* sp.) is a coniferous tree with over ninety different species. It is an important source of timber and industrial resin and hence it has been planted extensively in different parts of the world. In India pine trees are extensively grown in the Himalayan region at altitudes ranging from 1200-3300 m AMSL. While *P. roxburghii, P. longifolia, P. gerardiana, P. patula* and *P. excelsa* are distributed both in the N.W. and N.E. Indian Himalayas, *P. khasya* and *P. insularis* are distributed mainly in the N.E. region. Pine forests are found in abundance in the mid altitudes (1200-1800) of the Kumaon region of the N.W. Himalayas. Besides timber and resin, the pine forests contribute to the agro-ecosystem of the region in the form of plant nutrients, which are made available by the decomposition of the pine needle litter, The pine needle litter and macronutrient supplying potential from pine forests have been estimated at 6076 kg ha^{-1} and 534.6 kg ha^{-1} respectively (Singh and Bhatnagar, 1997). Due to this the decomposition of pine needles becomes important both from the point of recycling of nutrients in the forest floor and nutrient supply to agro-ecosystems.

But unlike broad-leaved litter pine needles are fairly recalcitrant to microbial decomposition owing to the presence of a wide range of inhibitory compounds that are present in the resin ducts of the pine needles. Hence the decomposition of pine needle litter has received much more attention from microbiologists, soil scientists and forest ecologists alike, the world over. In this chapter we will address the nature of pine needles and the various groups of microbes and enzymes that are involved in the decomposition of pine needle litter.

Composition of pine needle litter

Like any other coniferous litter pine needle litter is a ligno-cellulosic material with cellulose, hemicellulose and lignin as the major structural compounds. Besides, pine needles are characterized by the presence of other extractable compounds such as terpenes, resins

1. Scientist; 2. Principal Scientist, Vivekananda Institute of Hill Agriculture (ICAR), Almora - 263601, Uttarakhand, email: vpkas@nic.in

and phenols. The most important of the phenols are the tannins, which are one of the major inhibitory compounds present in pine needles.

The quality and quantity of the individual components may vary with the degree of maturity of the pine needles, the species being studied and the growing conditions. The litter compositions of the two major species *viz.*, *Pinus patula* and *P. roxburghii* are given in Table 1.

Table 1: Major litter composition of *Pinus patula* and *P. roxburghii* (unpublished data)

Component (%)	*Pinus patula*	*P. roxburghii*
Carbon	45.75	45.25
Nitrogen	0.435	0.46
C:N ratio	105: 1	98.35
Lignin	23.6	NA
Cellulose	37.2	NA
Hemicellulose	30.2	NA
Tannins NA 13.1		
(70% acetone extractable)		

NA–Not Analyzed

Microbes and enzymes involved in the decomposition of cellulose and hemicellulose

Cellulose, which is the most abundant structural compound on earth, is the major component of pine needles. Pine needles usually contain around 35-40% of cellulose and 30% hemicellulose. Cellulose is considered to be a linear polymer of hydroglucose units; linked with β-1-4-glucosidic units, although the true repeating stereochemical unit of cellulose is cellobiose (β-1-4-D-glucosyl-D-glucose). Fungi are the major group of organisms involved in the decomposition of cellulose. However relatively few organisms have the ability to produce all the necessary enzymes for the degradation of cellulose. While bacteria secrete cell bound enzymes, fungi are able to secrete these enzymes into the environment. Lists of cellulolytic microflora that are commonly encountered in the forest ecosystem are presented in Table 2.

Fungi causing soft rot mainly degrade polysaccharides. Among the soft rot fungi, *Trichoderma reesii* produces a complete set of endogluconases that includes at least three endogluconases, two exogluconases and one or two exogluconases and one or two β-glucosidases. Brown rot fungi degrade cellulose rapidly but their enzyme system seems to differ from the soft rot fungi. They produce endogloconases and β-glucosidases, but no detectable levels of exogluconases. The white rot group of fungi has in common the ability to degrade lignin as well. In the most studied white rot fungi *Phanerochete chrysosporium* at least five endogluconases, three exogluconases and two β-glucosidases have been characterized.

Both aerobic and anaerobic bacteria are known to degrade cellulose. However, the production levels of bacterial cellulases are quite low and are associated with the cells. The degradation of celluloses in bacterial systems seems to take place by a cluster of multienzyme complexes, which are difficult to disrupt without the loss of total activity as well as the individual components. The cellulosome of the well characterized bacterium *Cellulomonas* consists of several endogluconases, an exogluconase, a cellobiophosphorylase that breaks down cellobiose to glucose and glucose 1 phosphate, a cellodextrin phosphorylase which phosphorylases β-1-4 oligoglucans and two β glucosidases. Among the actinomycetes mesophilic species of *Streptomyces,* and thermophilic species of *Thermoactinomyces* and *Thermomonospora* are the most prevalent and their cellulosomes are similar to the bacterial systems.

Table 2: Certain important cellulolytic microflora involved in the decomposition of plant fibers in the forest ecosystem

Class	Microorganisms
Soft Rot Fungi	*Aspergillus niger*
	Chaetomium niger
	Fusarium oxysporum
	Neurospora crassa
	Penicillium pinophilum
	Trichoderma reesii
Brown Rot Fungi	*Coniophora putanea*
	Lanzites trabeum
	Poria placenta
	Tyromyces palustaris
White Rot Fungi	*Phanerochete chrysosporium*
	Sporotrichum thermophile
Aerobic Bacteria	*Bacillus circulans*
	Bacillus subtilis
	Cellulomonas firmi
	Cellvibrio gilvus
	Microbiospora bispora
	Pseudomonas fluorescens
Anaerobic Bacteria	*Acetovibrio cellulolyticus*
	Clostridium cellulovorans
	C. thermocellum
Actinomycetes	*Streptomyces lividans*
	Thermoactinomycetes curvata
	Thermomonospora fusca

Hemicelluloses are easily hydrolyzed short chains of branched heteropolysaccharides composed of both hexoses and pentoses. D-xylose and L-arabinose are the major constituents of the pentosans (xylans), while D-glucose, D-galactose and D-mannose are the constituents of the hexosans (mannans). Hemicellulases are wide spread in nature and the hemicelluloses of fungal origin are the most studied. Among the fungi the most prominent producers of hemicellulase are *Trichoderma, Aspergillus. Fusarium,* and several brown rot fungi such as *Tyromyces palustris, Poria placenta* and *Coniophora cerebella.* Hemicellulases of bacterial origin are less studied and are confined to bacteria of the genus *Bacillus, Streptomyces, Cellulomonas, Thermomonospora* and *Chainia* sp. The major anaerobic bacteria producing hemicellulases are *Clostridium, Thermoanaerobacter, Acetovibrio* and *Bacleroides* (Table 3).

Table 3: Important sources of hemicellulases involved in the decomposition of plant fibers in the forest ecosystem

Enzyme	Bacteria	Fungi
β-1-4-Xylanases	*Bacillus pumilis*	*Aspergillus niger*
	B. subtilis	*Fusarium oxysporum*
	Clostridium acetobutylicum	*Aspergillus wentii*
	Cellulomonas uda	*Trichoderma koningii*
	Streptomyces xylophagus	*Neurospora crassa*
β-l-4-Xylosidases	*Clostridium thermocellum*	*Aspergillus niger*
	Bacillus pumilis	*Corticum rolfsii*
	Acetovibrio cellulolyticus	*Penicillium wortmanii*
		Trichoderma reesii
α- Arabinosidase	*Streptomyces purparescens*	*Aspergillus niger*
	Bacillus subtilis	*Cortisium rolfsii*
	Ruminococcus albus	*Trichoderma reesii*
α-Glucuronidase		*Trichoderma reesii*
		Agaricus bisporus
		Pleurotus osteoratus
Esterases	*Fibrobacter sucinogenes*	*Aspergillus niger*
	Bacteroides cellulosolvens	*A. phoenicus*
	Clostridium thermocellum	*Tricoderma reesii*
β-l-4-Mananase	*Aerobacter aerogenes*	*Aspergillus niger*
	Bacillus subtilis	*Thielavia terrestris*
	Caldocellum saccharolyticum	*Trichoderma reesii*
	Streptomyces lividans	*Paecilomyces variotii*
β-1-4-Mannosidases	*Bacillus subtilis*	*Aspergillus niger*
		A. awamorii
		Thilavia terrestris
		Polyporus sulphureus
α-Galactosidase	*Bacillus subtilis*	*Aspergillus niger*
		Sclerotium rolfsii
		Aspergillus tamari
		Mortierella vinacea

Decomposing pine needles are colonized mainly by *Trichoderma* and *Aspergillus* (Selvakumar unpublished data), and they seem to play the role of "sugar fungi" which specialize in the utilization of cellulose and hemicellulose, in case of pine needle decomposition.

Microbes and enzymes involved in the decomposition of lignin

Lignin is not a definitive uniform compound, but is a collective form for substances that have similar chemical properties but very different molecular weights. Pine needle are a rich source of lignins and the lignin concentration varies from 27-30% depending on the species. The molecular weight of lignins may be 100 kDa or greater. A considerable part of the photosynthetic activity of plants is devoted for the synthesis of lignin. Lignins are highly branched polymeric molecules consisting of phenyl- propane base monomeric units linked together by different types of bonds including alkyl-aryl, alkyl-alkyl, and aryl-aryl bonds. The relative proportion of the three cinnamyl alcohol precursors incorporated into lignin *i.e.* p-coumaryl alcohol, coniferyl alcohol and sinapyl alcohol vary with the plant species involved. Gymnosperm lignin characteristically contains a preponderance of the vanillyl phenylpropanoid moiety. In terms of decomposability by the white rot group of fungi, which are the major decomposers of pine needle litter, the vanillyl moiety has a significantly lower rate of decomposition than the other moieties.

Lignin decomposition takes place in nature mainly by the action of fungi, of which the white rot group of fungi is the most important. Most of the white rot group of fungi belongs to Basidiomycetes except for *Ustilinia vulgaris,* which is classified as an ascomycete. These fungi simultaneously attack the polysaccharides and lignin by their complex enzyme systems. The degradation of lignin by white rot fungi is oxidative in nature and three phenol oxidases are found to play an important role in the lignolytic system *viz.,* laccase, lignin peroxidase (LiP) and manganese dependent peroxidase (MnP). Laccase is a true phenoloxidase that has broad substrate specificity. It oxidizes phenols and phenolic lignin substances with formation of radicals. The enzyme can cause both polymerization and depolymerization by the production of phenoxy radicals. However laccase is generally considered to have a redox potential too low to allow for the oxidation of non-phenolic lignin structures. The various groups of lignolytic enzyme producing while rot fungi are detailed in Table 4.

Bacterial degradation of lignin has been less studied and a few Gram-positive bacteria belonging to the genera *Pseudomonas, Xanthomonas, Acinetobacter, Achromobacter, Agrobacterium. Aerobacter* and *Erwinia* have been reported to degrade lignin. The nature and function of the bacterial enzyme systems are much different from the fungal systems and most of these enzymes are intracellular in nature. Actinomycetes degrade grass lignins by the release of lignin rich water-soluble fragments called Acid Preciptable Polymeric Lignin (APPL). These fragments are not metabolized further by the actinomycetes and APPL production seems to be mainly restricted to the genus *Streptomyces* and *Thermomonospora.* With the discovery of LiP from *Streptomyces* spp., it seems likely that these actinomycetes attack lignin by a similar method to that of white rot fungi. While the role of several white rot fungi in the decomposition of pine needle litter is well established in the role of bacteria and actinomycetes is yet to be established. The initial phase of lignin degradation is thought to involve phenyloxidase and laccase enzymes (Ander *et al.*, 1990).

Table 4: Major groups of white rot fungi and their enzymes

LiP, MnP and laccase producing	*Coriolus versicolor*
	Cyathus bulleri
	Phelbia radaita
	Pycnoporus sanguineus
	Pleurotus osteoratus
	P. sajor-caju
LiP and MnP producing	*Coriolus pruniosum*
	Phanerochete chrysosporium
MnP and laccase producing	*Ceriporiopsis subvermispora*
	Dichomitus squalens
	Lentinus edodes
	Pinus tigrinus
	Rigidioporus lignosus
LiP and laccase producing	*Pleurotus florida*
	P. pletensis
	P. brumensis
	Phlebia tremellosus
	P. ochracefolia
LiP producing	*Bjekandera adusuta*
	Polyporus varius
	Daedaleopsis confragosa
Laccase producing	*Pycnoporus cinabarinus*
Novel oxidase producing	*Pleurotus eryngii*
	P. ostearatus
	Trametes versicolor
	Pleurotus sajor-caju

LiP = Ligrin peroxidase
MnP = Manganase dependent peroxidase

Importance and effect of tannins on pine needle litter decomposition

Estimated to be the fourth most abundant biochemical produced by vascular plant tissue after cellulose, hemicellulose and lignins, tannins represent a significant portion of terrestrial biomass. In woody species foliar tannin levels commonly range from 15-25% dry weight but concentrations as high as 40% have been reported. Tannins enter the soil by leaching and litter inputs both above and below ground. A large fraction of tannins and simpler phenolics appear to enter the soil as foliar leachate. Tannins are commonly defined as water-soluble polyphenolic compounds ranging in molecular weight from 500 to 3000 Daltons that have the ability to precipitate proteins (Bale-Smith, 1977). Tannins, which are present in most plant systems, are meant to be defence systems against microbial attack and foraging by animals. Tannins have been reported to be inhibitors of several microbial and enzymatic activities (Donovan and Brooker, 2001). They form weak pH dependent associations with a range of substrates including cellulose, proteins enzymes, fats, nucleic and amino acids, often making the substrate resistant to microbial attack.

Tannins are of two types *viz.*, hydrolysable and condensed. The hydrolysable tannins are polyphenols with a polyol central core, while the condensed tannins are polymers of flavon-3-ol, which are linked by carbon - carbon bonds. Gymnosperms like pine are known to produce only condensed tannins while dicots are known to produce both hydrolysable and condensed tannins. The pine needles are a rich source of both hydrolysable and condensed tannins, which inhibit the decomposition process to a great extent (Northup *et al.*, 1995).

Tannin have demonstrated intermediate lability compared to the other major biochemical components. Studies on decomposing *P. radiata* needles suggested that the concentration of condensed tannins was 12-22%. Changes in the ^{13}C NMR spectra of tannins of decomposing *Pinus radiata* needles suggested that chemical alteration of the tannins occurred during the first 4-12 months of decomposition but no further chemical alteration was detected during the fallowing 2-3 years (Parfitt and Newman, 2000, Table 5).

Table 5. Effect of different pine needle tannin extracts on some beneficial fungi

Tannin Concentration*	*Trichoderma virescens* (MTCC 6321)		*Trichoderma reesii*		*Pleurotus djamor*	
	Zone of Inhibition (mm)					
	Water Extract	Acetone Extract	Water Extract	Acetone Extract	Water Extract	Acetone Extract
100 ppm	NI	NI	NI	NI	NI	NI
500 ppm	NI	NI	NI	NI	NI	NI
1000 ppm	NI	NI	NI	NI	NI	NI
2000 ppm	NI	NI	NI	NI	NI	NI
3000 ppm	NI	NI	NI	NI	NI	NI
4000 ppm	32.9	30.0	30.0	25.0	NI	NI
5000 ppm	37.0	33.5	32.6	30.5	NI	NI
10000 ppm	64.5	59.2	60.0	58.7	NI	NI

(Selvakumar, unpublished data)

NI–No Inhibition; * - Tannic Acid Equivalent

Tannins like lignin and other polyphenols are resistant to decomposition. Some white rot fungi like *Sporotrichum pulverulentum, Ceriporiopis subvermispora, Cyathus steroreus, Coriolopis gallica* are able to degrade tannins. Other fungi capable of growing on tannins as sole C or that can degrade tannins include the genera *Aspergillus, Fusarium, Cylindrocarpon, Gliocladium, Endothia, Calvatia, Penicillium* and *Trichoderma*. There is less evidence demonstrating bacterial degradation of tannins although bacteria can tolerate high concentrations of tannins (Gamble *et al.*, 1996). Bacterial species known to use tannin as the sole C source include *Achromobacter, Azotobacter, Bacillus, Corynebacterium, Escherichia, Klebsiella, Pseudomonas,* and *Staphylococcus.* Tannase an enzyme that catalyses the hydrolysis of the ester between phenolic acid and an alcohol, has been reported in fungi, bacteria and yeast. Polyphenol oxidase is capable of polymerizing and inactivating tannins. Tannin decarboxylation can be carried out by phenol oxidases and white rot fungi release these exoenzymes.

We have observed that crude extracts of water and solvent extractable tannin fractions from needles of *Pinus roxburghii* were found to contain tannin concentrations of 10.15% and 13.15% in terms of tannic acid equivalents respectively. Both the fractions were found to possess inhibitory activity against several microbes of agricultural importance. Among the fungi studied we observed that *Pleurotus djamor* was found to and tolerate up to 10000 ppm of crude tannin concentration, while both *Trichoderma virescens* (MTCC 6321) and *T. reesii* were inhibited by crude tannin concentrations above 3000 ppm.

Microbial succession in decomposing pine needle litter

In the context of conifer litter, decomposition is usually initiated by primary colonizers involving a diverse community of fungi and bacteria, which utilize simple sugars, oligosacharides and other low molecular weight compounds. After this initial flush of microbial activity specialist secondary colonizers notably the basidiomycetes that are less competitive than the microfungi in exploiting labile resources effect the decomposition of recalcitrant plant polymers such as lignocellulosic complexes. Exceptions to this paradigm occur when the time course of fungal succession is attenuated by dominant lignolytic basidiomycetes such as *Mycena gallopus* and *Marasmius androsarceus* invading from established mycelial networks (Frankland, 1992).

Certain fungi are found to be dominant in decomposing pine needle litter. Cox *et al.*(1997) observed that in pine needle litter that was inoculated with *Trichoderma,* no other species of fungi was isolated after six months of incubation. In one of the studies Cox *et al.* (2001) investigated the survival of two common fungal species *Marasmius androsaceus* and *Trichoderma viride* on sterile pine needle litter in litterbags placed in the field. *M. androsaceus* is a strongly lignolytic and cellulolytic species while *T. viride* can actively decompose cellulose. They were able to observe that in litter inoculated with *T. viride* no other fungi could be isolated over 12 months, while after 12 months *Acremonium* spp. and *Moritella* sp. were also present. There was no *M. androsaceus* recorded in the litter after 2-3 months and no hyphae with clamp connections were observed on needles in this treatment between 3-6 months. However the same karyotype was isolated from the final samples of *Marasmius* litter at 12 months indicating its presence in or in the vicinity of the litter. In total 26 species of common hypomycetes and deuteromycetes were isolated from those two treatments. Most of the fungi were cellulolytic but not lignolytic. It was observed that the litter inoculated with *M.androsarceus* lost 17% of the initial lignin over the first six months and 38% over the last six months while cellulose decomposition was 15% at 6 months and 51% at 12 months. In contrast litter inoculated with *Trichoderma* showed no significant reduction in initial mass of lignin over 12 months while cellulose degradation was 33% and 62% at 6 and 12 months respectively. The uninoculated control treatments recorded cellulose losses of 26 and 62% at 6 and 12 months respectively. It was concluded from this study that the process of decomposition in litter inoculated with *Trichoderma* were similar to those in litter that was naturally colonized by several species of fungi. This suggests that this single dominant species of fungus was very effective in the decomposition of structural carbohydrates in pine litter as a complex naturally occurring fungal assemblage. The absence of *Marasmius* from the inoculated litter after a few months was attributed to the poor ability of many basidiomycetes to compete with the sugar fungi until the simple C compounds are depleted. The lignin degradation occurred progressively in the *M. androsaceus* treatments. This suggests that either *Marasmius* initially affected the biochemical structure of lignin so that it could be affected by the non-lignolytic species.

Factors affecting pine needle decomposition

a. Nitrogen

Nutrients are hypothesized to limit rates of litter decomposition in natural ecosystems. The decomposability of litter inputs to soils their C: N ratio and lignin content is well established. Immobilization of nutrients takes place when litter of poor quality is decomposed. Though in general the addition of nitrogen is known to increase the rate of decomposition. In lignin rich environments like pine needles the addition of N may inhibit the decomposition of lignin either by suppressing the synthesis of lignolytic enzymes or by promoting the formation of additional recalcitrant compounds (Berg and Matzner, 1997). Usually it is observed that the pine needle litter has a high C: N ratio compared to the needles. This may be attributed to the leaching of labile N constituents from needles upon fall.

Boyle (1998) while testing the effect of addition of different N containing compounds in a laboratory wood decomposing experiment showed that the inhibition of lignin degradation was higher with simple N sources than with complex sources. Virzo De Santo *et al.* (2002) observed that after 6 months of decomposition basidiomycetes reached a peak on N poor pine needles but were absent from N rich pine needles of *Abies alba.* It has been hypothesized that the initial fast decomposition would increase the N concentration, which would inhibit the growth of many lignin-degrading fungi.

b. Climatic conditions

Litter decomposition being a predominantly microbial process, it is largely regulated by the external environment *viz.*, the temperature conditions and the precipitation. It has been hypothesized that wet and warm climate would favour fast growing microbes in the early stages of decomposition amongst the less competitive fungi especially the basidiomycetes. Several studies have shown that while holocellulose degradation and N release were controlled by temperature cellulose degradation was largely unaffected by the temperature. Couteaux *et al.* (2002) found that *Pinus halepense* litter incubated at temperatures ranging from 23-30°C at optimal moisture after 7 months 50% of the initial cellulose was lost independent of the temperature. In contrast to cellulose the rate of hemicellulose appeared to have been more dependent on temperature regimes. The correlations with temperature dependent climate variables indicate that the hemicellulose mass loss increased with increasing temperature.

Precipitation affects pine needles mainly by controlling the water availability for microbial activity and by the leaching of the nonstructural compounds present in the needle litter. McTiernan (2003) found a very strong relation between the precipitation and cellulose decomposition, and lignin increase in the decomposing pine needle litter. Lignin concentrations in decomposing pine needle litter were found to be significantly correlated to the mean annual precipitation and water surplus. He further observed that the decomposition of pine needles was much more sensitive to water deficit, in comparison to spruce litter, due to the arrangement of pine needles in pairs, resulting in a litter layer with large air spaces and hence more moisture loss from the soil surface.

c. Growth conditions of pine trees

The decomposability of litter inputs to soils their C to N ratio and lignin content is well established. In the temperate forest ecosystems there is a clear and more complex two phase system of litter decomposition described by exponential mass loss wherein the rapid decomposition of a labile fraction is followed by the slow decomposition of a more recalcitrant less biodegradable fraction related to lignin concentration. The quality, quantity and spatial configuration of the lignocellulosic complex in the plant cell wall is therefore a key determinant of litter mass loss while concentrations of labile carbon and nutrients (particularly N and P) influence the development of microbial biomass exploiting the resource.

The biochemical composition of forest litter is primarily a function of the tree species but intraspecific responses to soil conditions and physical environmental factors also result in considerable variation in foliage and litter quality. In soils of low and average fertility trees have low foliar nutrient, concentration higher reabsorption of N and P before leaf abscission than those growing on more fertile soils. Higher concentrations of polyphenols are found in foliage of trees under stress. Several authors have noticed considerable differences in litter constituents within the same species in different soils, which can affect decomposition rates. Berg *et al.* (1993) observed an inverse relationship between the soil N availability and N concentration of *Pinus sylvestris* needles. Sanger *et al.* (1996) showed that the amount, composition and degree of lignin polymerization varied between soils of different base status. But very few studies have focused on this important parameter and there is a need to clearly identify the association between plant litter qualities with incident soil characteristics.

Role of pine needle litter in the N.W. Himalayan agricultural ecosystem

In the low external input agricultural system of the N.W. Himalayas, pine needle based compost is one of the most important inputs in crop production. The local farmers for compost production collect the pine needles that fall in the months of April-May. The litter is usually heaped outside and exposed to the natural elements till winter when they are used as bedding material for cattle. This bedding material soaked in cattle dung and urine is usually used for manuring the fields during the next spring and *Kharif* seasons. Most often it is observed that the pine needle compost used by the farmers is partially decomposed with large chunks of visible partially decomposed pine needles. This compost is also found to be of very poor quality and little is known about the use of such compost on the beneficial soil microflora.

Besides compost, pine needles components also enter the agro-ecosystem as runoff from pine forests. The topography of the region is such that the runoff from the pine forests usually enters the cultivated terraced land situated below the pine forests thus bringing with it large quantities of tannin rich leachates from the pine forest floor and pine needle litter. The role of pine needle decomposition products and secondary compounds on the soil microbial ecosystem and their allelopathic effects on plants needs to be studied extensively.

Conclusion

Pine is an economically important tree of the temperate regions of the world. Due to this fact it has been planted in a large scale in the Himalayan region of India. It forms an integral part of the agro ecosystem of the region and the needle litter constituents enter the agro-ecosystem in many ways. The needle decomposition pattern also affects the nutrient cycling pattern in the forest ecosystem and is largely responsible for the regulation of the soil microbial communities and their activities. Though many studies have been conducted especially in the Scandinavian nations with regard to the needle litter decomposition and nutrient release pattern from decomposing needle litter, most of the literature deal mainly with the decomposition patterns under naturally occurring forest conditions with little emphasis on the microbial communities that make up the complex web of decomposition. The succession pattern of naturally occurring microbial communities in the ecosystem and the effect of inoculation of a particular microbe/consortia on the decomposition pattern need to be studied thoroughly.

From the practicality of the problem it would be useful if a suitable microbial consortia were developed for hastening the decomposition process of pine needle litter, which has become an important input in Himalayan agriculture. Alternatively suitable decomposition approaches such as vermicomposting can be evolved so as to hasten the process of decomposition at the farmers level at a low cost. Such a strategy should focus on the mitigation of the negative effects of the different inhibitory compounds such as tannins that are plentiful in needle litter.

References

Ander, P., Chitrra, M., Roberta, L.F. and Erikkson, K.E. 1990. Redox reactions in lignin degradation: interactions between laccase, different peroxidases and cellobiose: quinone oxidoreductase. *Journal of Biotechnology,* **13**: 189-198.

Bate-Smith, E.C. 1977. Astringent tannins of *Acer species. Phytochemistry,* **16**: 1421-1426.

Berg, B. and Matzner, E. 1997. Effect of N deposition on decomposition of plant litter and soil organic matter in forest ecosystems. *Environ. Rev.,* **5**: 1-25.

Berg, B., McClaugherty, C. and Johansson, M.B. 1993. Litter mass loss rates in late stages of decomposition at some climatically and nutritionally different pine sites. Long-term decomposition in a Scots pine forest. *Canadian Journal of Botany,* **71**: 680-692.

Boyle, D. 1998. Nutritional factors limiting the growth of *Lentinula edodes* and other white rot fungi in wood. *Soil Biology* and *Biochemistry,* **30**: 817-823.

Couteaux, M.M., Aloui, A. and Kurz-Besson, C. 2002. Temperature and faunal controls on *Pinus halepense* litter decomposition as influenced by temperature and a millipede *Glomeris marginata*. *Applied Soil Ecology,* **20**: 85-96.

Cox, P.J., Fischer, J. and Anderson, J.M. 1997. Experiments in fungal survival of two common pine needle colonizers. *Mycologist,* **11**: 55-58.

Cox, P., Wilkinson, S.P. and Anderson, J.M. 2001. Effects of fungal inocula on the decomposition of lignin and structural polysaccharides in *Pinus sylvestris* litter. *Biol. Fertil. Soils,* **33**: 246-251.

Donovan, L.O. and Brooker, J.D. 2001. Effect of hydrolysable and condensed tannins on growth morphology and metabolism of *Streptococcus gallolyticus* and *Streptococcus bovis. Microbiology,* **147**: 1025-1033.

Dutta, D.K., Banerjee, S.K. and Gupta, S.K. 2001. Decomposition of broad-leafed and coniferous forest litter. *Journal of the Indian Society of Soil Science,* **49**: 496-499.

Frankland, J.C. 1992. Mechanisms in fungal succession. In: *The Fungal Community* (eds.) G.C. Caroll and D.T. Wicklow. Dekker, New York, pp. 340-383.

Gamble, G.R., Akin, D.E. and Makkar, H.P.S. 1996. Biological degradation of tannins in *Sericea lespezeda* (*Lespezeda cuneata*) by the white rot fungi *Ceriporiopsis subvermispora* and *Cyathus stercoreus* analyzed by solid state ^{13}C nuclear magnetic resonance spectroscopy. *Appl. Environ. Microbiol.* **62**: 3600-3604.

McTiernan, K.B., Couteaux, M.M., Berg, B., Berg, M.P., Rosa Calvo de Anta, Gallardo, A., Kratz, W., Piussi, P., Remacle, J. and Virzo De Santo, A. 2003. Changes in chemical decomposition of *Pinus sylvestris* needle litter during decomposition along a European coniferous forest climatic transect. *Soil Biology and Biochemistry.* **35**: 801-812.

Northup, R.R., Yu, Z., Dalgren, R.A. and Vogt, K.A. 1995. Polyphenol control of nitrogen release from pine litter. *Nature.* **377**: 227-229.

Parfitt, R.L. and Newman, R.H. 2000. ^{13}C NMR study of pine needle decomposition. *Plant Soil.* **219**: 273-278.

Sanger, L.J., Cox, P., Splatt, P., Whelan, M.J. and Anderson, J.M. 1996. Variability in the quality and potential decomposability of *Pinus sylvestris* litter from sites with different soil characteristics. I. Lignin and phenyl propanoid structures. *Soil Biology and Biochemistry.* **28**: 829-835.

Singh, R.D. and Bhatnagar, V.K. 1997. Differences in soil and leaf litter nutrient status under pine, cedrus and quercus. *Indian Journal of Forestry.* **20**: 147-149.

Virzo De Santo, A., Rutiglianao, F.A., Berg, B., Fioretti, A., Puppi, G. and Alfani, A. 2002. Fungal mycelium and decomposition of needle litter in three contrasting coniferous forests. *Acta Oecologia,* **23**: 247-259.

Soil Microbiology and Soil Biotechnology : Vol. 1
Editor : S. Kannaiyan
Associated Publishing Co., New Delhi, 2010, pp. 43-62

4

Nitrification in Soil

Balraj, S. Parmar[1] *and Rajesh Kumar*[2]

Introduction

Nitrogen (N) is essential for all life. It is an essential element for plant growth and reproduction and is one of the most widely distributed elements in nature, with atmosphere as the main reservoir. Of the total N found in nature, 99.96% is present in the atmosphere. Of the remaining 0.04%, biosphere contains only 0.005%. In spite of the small proportion found in living beings, N is most often the limiting nutrient for crop production since only a fraction of atmospheric nitrogen is made available to the plants through biological nitrogen fixation. However, introduction and use of chemical nitrogenous fertilizers has resulted in substantial increase in crop yields. These fertilizers have made a major contribution in improving agricultural productivity the world over. Ammonium based fertilizers including urea are the extensively used source of fertilizer N for field crops. However, fertilizer N is not an unmixed blessing as the commonly used nitrogenous fertilizers, especially urea, suffer from low nitrogen use efficiency (NUE) and contribute towards environmental pollution and health hazards. Worldwide, the NUE for cereal production (wheat, corn, rice, barley, sorghum, millet, oat, rye etc.) is approximately 33 per cent. The unaccounted 67% represents a US $15.9 billion annual loss, assuming fertilizer-soil equilibrium (Raun and Johnson 1999).

The soil accounts for a minute fraction of the lithospheric N. It is, however, the major source of plant-available N. Out of the total N available in soil, only a small fraction is available to plants as ammonium and/or nitrate. Although most plants are able to use both ammonium-N and nitrate-N, the latter is the predominant form consumed by the plants under aerable/terrestrial situations. This is primarily because of compulsion rather than preference as most of the ammonium-N is rapidly nitrified under optimum conditions of temperature, moisture and aeration. As a result, ammonium-N is available to the plants only for a limited period of time. Most of the time, plant roots come across nitrate-N as a source of N.

1. Joint Director (Research), Indian Agricultural Research Institute, New Delhi-110 012; bsparmar@yahoo.com
2. Scientist (Sr. Scale), Division of Agricultural Chemicals, Indian Agricultural Research Institute, New Delhi-110 012; rkb_1973@yahoo.co.in

Nitrification is a key process in managed agricultural ecosystems because the conversion of ammonium to nitrate can lead to substantial loss of agricultural N (both applied as chemical fertilizers or present as native soil N) by leaching and/or denitrification. The loss ol fertilizer N is of concern not only because of economic reasons but also due to the pollution potential of different N forms. Some of the adverse environmental effects of excessive use of nitrogenous fertilizers include i) methemoglobonemia ("Blue Baby" Syndrome) in infants due to nitrate and nitrite in waters and food, ii) cancer due to secondary amines, iii) respiratory illness due to nitrate, nitrite and HNO_3 in aerosols, iv) eutrophication due to N in surface waters, v) material and ecosystem damage due to HNO_3 in rain water, vi) plant toxicity due to high levels of nitrite and ammonium in soils, vii) excessive plant growth due to more available N, and viii) depletion of stratospheric ozone due to NO and N_2O. It is because of these concerns that concerted efforts have been and are being made to reduce the use or increase the efficiency of fertilizer N uptake by crop plants (Prasad and Power 1995; Prasad 1998; Azam and Farooq 2003).

Multi-disciplinary approaches have been identified and put to practice over the past few decades to increase use efficiency and decrease the losses of fertilizer N. These include a) breeding crop varieties with higher fertilizer use efficiency, b) management of fertilizer N, c) use of chemicals inhibitory to specific N transformation processes like urea hydrolysis and nitrification *i.e.*, urease and nitrification inhibitors, respectively, to slow the process of ammonium formation from urea and to regulate the process of nitrate formation, d) supplementation/integration of fertilizer N with organic manures *e.g.*, green manures. These approaches applied singly or in combination, have certainly helped alleviate the problems arising as a result of fertilizer N use. Fertilizer management through improved formulations, mode and time of application and placement etc. has also been found helpful to overcome some of the problems. Likewise, a large variety of chemicals have been tested as potent inhibitors of specific N transformation processes including urea hydrolysis and nitrification. With these inhibitors, not only use efficiency of fertilizer N is enhanced, but significant reduction occurs in losses due to ammonia volatilization, denitrification and nitrate leaching.

Amongst different N cycle processes vis-a-vis ecosystem functioning and environmental concerns, the process of nitrification stands out to be the most important process. Nitrification inhibition could lead to: i) increased rhizospheric microbial activities including biological nitrogen fixation and production of phytohormones, ii) enhanced mineralization of native soil N thus decreasing dependence on chemical fertilizers, iii) increased efficiency of fertilizer N through decreased losses via denitrification and nitrate leaching and iv) greater photosynthate partitioning to the rhizosphere thus enriching the soil with organic matter. It is important, therefore, to develop an understanding of nitrification process, factors affecting nitrification, methods to regulate the process and its implications to ecosystem functioning.

The nitrification process

Nitrification has been defined as the oxidation of any reduced nitrogen form (organic or inorganic) to nitrate. The microorganisms carry out this oxidation process. Being the only link between reduced and oxidized nitrogen compounds, the nitrification process is of major importance for the nitrogen cycle in aquatic and terrestrial environments. In soils, the nitrification process oxidizes the immobile ammonium to nitrate, a mobile ion. The biological oxidation of ammonia to nitrate is a two-step process mediated by autotrophic bacteria. It is first oxidized to nitrite and then to nitrate, as follows:

$$NH_3 + O_2 \rightarrow NO_2^- + 3H^+ + 2e^-$$

$$NO_2^- + H_2O \rightarrow NO_3^- + 2H^+ + 2e^-$$

The source of ammonium-N could be soil organic matter (mineralization by soil microorganisms) and/or chemical fertilizers. In the case of soil, organic N is used by the ammonifiers, while chemical fertilizers either contain ammonium as such or its precursors.

The oxidation of reduced organic or inorganic nitrogen to nitrate mediated by heterotrophic organisms has been named heterotrophic nitrification, whereas the oxidation of reduced inorganic nitrogen to nitrate by autotrophic organisms is called autotrophic nitrification.

Several genera and species of ammonium and nitrite oxidizing heterotrophs including fungi (*Aspergillus flavus, Neurospora crassa, Penicillium* sp.), actinomycetes (*Streptomycetes* sp., *Nocardia* sp.) and bacteria (*Arthrobacter* sp., *Azotobacter* sp., *Pseudomonas fluorescens, Aerobacter aerogenes, Bacillus megaterium, Proteus* sp.) have been reported. However, autotrophic nitrifiers are generally considered as the main organisms responsible for most of nitrification. Ammonium oxidizing autotrophs include *Nitrosomonas, Nitrosolobus* and *Nitrosospira*. These organisms have been isolated from a variety of soil environments with ubiquitous distribution. Nitrite produced by the ammonium oxidizing autotrophs is rapidly oxidized to nitrate by *Nitrobacter* species.

All nitrifiers are obligate aerobes and hence a restricted nitrification under waterlogged or aquatic environments can be observed. In addition, these microorganisms, especially *Nitrobacter,* are fairly sensitive to acidic pH. As a result nitrification is inhibited in climax ecosystems like forest soils with thick layer of leaf litter and zones of acidic pH. The process of nitrification itself may lead to lowering of pH of the medium due to release of H^+ as shown in the equation above.

As stated earlier, the autotrophic bacteria mostly carry out the nitrification process. These utilize reduced inorganic nitrogen as energy source, and carbon dioxide as carbon source. Hence, this nitrification process may more correctly be called chemolithoautotrophic nitrification. The substrate for the enzyme ammonia monooxygenase (AMO) involved in the first part of the chemolithoautotrophic nitrification process is ammonia rather than ammonium but an acidity dependent equilibrium always exists between ammonia and ammonium. It is mediated by two distinct groups of bacteria, ammonia-oxidizing bacteria (AOB) and nitrite-oxidizing bacteria. Both groups are dominated by autotrophic metabolism; however, some nitrite oxidizers may also use organic compounds as carbon sources (mixotrophs).

Chemolithoautotrophic ammonia-oxidizing bacteria (AOB)

The AOB carry out a specific environmental function of oxidation of ammonia to nitrite. All AOB that have been isolated and characterized are gram-negative, obligate aerobic and obligate chemolithoautotrophs. Together with the nitrite-oxidizing bacteria, they make up the family *Niirobacteraceae.* Five different genera for AOB were defined on the basis of classical morphological characteristics: *Nitrosomonas, Nitrosospira, Nitrosococcus, Nitrosolobus* and *Nitrosovibrio.* Most AOB described belong to the β-*Proteobacteria* (β-AOB) but a few marine isolates of *Nitrosococcus,* not known from soil, belong to the γ-*Proteobacteria* (γ-AOB). Recent research based on sequences of 16S rDNA suggest that the β-AOB can be divided into only two major phylogenetic lineages, the *Nitrosospira* and the *Nitrosomonas* lineages. New results showed similar, but not identical, evolutionary relationships of β-AOB when using the 16S rRNA gene or a functional gene *(amoA,* see below) as marker genes for phylogenetic analysis (Purkhold *et al.,* 2000). Hence, except for a few marine *Nitrosococcus* strains, all known AOB are of monophyletic origin. The high correlation between function and phylogeny is rather unique for AOB when compared to other functional groups of microorganisms in soil, *e.g.* denitrifiers (Bothe *et al.,* 2000).

Two key enzymes mediate the ammonia oxidation in AOB: ammonia monooxygenase (AMO) and hydroxylamine oxidoreductase (HAO) and both are co-dependent because they generate substrate and electrons, respectively, for each other (Bothe *et al.,* 2000).

Oxidation of ammonia to hydroxylamine (AMO):

$$NH_3 + O_2 + 2H^+ + 2e^- \rightarrow NH_2OH + H_2O$$

Oxidation of hydroxylamine to nitrite (HAO):

$$NH_2OH + H_2O \rightarrow NO_2^- + 5H^+ + 4e^-$$

The AMO enzyme consists of 3 subunits with different sizes: AMO-A, AMO-B and AMO-C (Bothe *et al.,* 2000), and mainly the gene (*amo-A*) encoding the A subunit, which carry the active site of AMO, has been investigated (Purkhold *et al.,* 2000). The AMO enzyme may catalyze co-oxidation of a broad range of substrates (McCarty, 1999). Hence, this enzyme has been in focus when exploiting the role of AOB in bioremediation (Duddleston *et al.,* 2000). Many similarities between AMO of AOB and paticulate methane monooxygenase (pMMO) found in methane-oxidizing bacteria have been reported (Bedard and Knowles, 1989) and the similarity of nucleotide sequences encoding the enzymes indicate a common evolutionary origin (Holmes *et al.,* 1995).

Inhibition or regulation of nitrification

As mentioned above, nitrification is one of the key N cycle processes under most arable situations on land. The following discussion will highlight the need to inhibit nitrification in order to maintain the economy of agro-ecosystems.

Nitrification as a source of atmospheric N_2O

Concentrations of N_2O in the atmosphere have increased markedly in recent years and are continuing to do so at 0.25% per annum. This increase is attributed mainly to biospheric processes. Flood irrigation leads to rapid nitrification/denitrification resulting in considerable amounts of N_2O being emitted to the atmosphere, which may amount to 35-45% of the applied N. On the global level, >65% of the atmospheric N_2O comes from the soil, which is twice the amount produced by burning fossil fuels and four times the amount evolved from the oceans. Rates of denitrification in soil may be as high as 30 kg N lost ha^{-1} day^{-1}, more realistically being around 3 kg N ha^{-1} day^{-1}. Being a greenhouse gas, N_2O contributes substantially to the destruction of stratospheric ozone (Azam and Farooq, 2003). Efforts have therefore been devoted to identify the processes that contribute to N_2O emissions and possible mitigation strategies.

Among different processes, nitrification and denitrification are reported to be the main contributors to atmospheric N_2O (Azam *et al.*, 2002). However, since the two processes occur simultaneously (aerobic and anaerobic microsites can develop within the same aggregate, supporting nitrification and denitrification, respectively), it is not easy to ascertain the real contribution of either process to the observed N_2O fluxes. Nevertheless, nitrification is reported to make a substantial contribution to the N_2O emission under aerobic conditions. Higher N_2O emissions are often reported from fertilized (a routine practice in the present-day agriculture) than unfertilized soils, rates of emission being greatest following application of ammonium or ammonium-forming fertilizers (Azam *et al.*, 2002). In several studies, using isotope methodology and nitrification inhibitors, this increase is attributed to losses of N_2O occurring during the process of nitrification (Abbasi and Adams, 2000). Estimates of the amount of N_2O resulting from nitrification are variable but generally account for < 1% of the fertilizer N applied. In the case of anhydrous NH3, however, the losses may increase to 6-7%. In most studies, the onset of N_2O emission is observed very early during the incubation, while nitrification continues for extended periods of time. Williams *et al.* (1998) reported active nitrification 7-12 days after application of ammonium nitrate, while a flush of N_2O emission from soil was observed around day 1, followed by a decline. These authors showed very low molar ratios of NO to N_2O and suggested that denitrification was the dominant process involved in N_2O emission. Azam *et al.* (2002) have reached a similar conclusion from model experiments conducted under laboratory conditions.

Under the semi-arid agro-climatic conditions and urea as major N fertilizer, contribution of nitrification to N_2O emissions may be higher than generally reported. Urea is rapidly hydrolyzed followed by a quick nitrification of the resultant ammonium especially when applied under relatively warmer conditions. Thus not only nitrification contributes to N_2O emissions but the process of denitrification is fairly well supported by sustained availability of nitrate. Because of the low organic matter content of the soils that would limit the process of denitrification, a major source of N_2O evolved may still be nitrification. In most soils, formation and emissions of N_2O to the atmosphere are enhanced by an increase in available mineral N through increased rates of nitrification and denitrification. Therefore addition of N in organic or inorganic compounds eventually leads to enhanced N_2O emissions.

Pollution of ground water

Uncontrolled and excessive nitrification may lead to ground water contamination with nitrate and nitrite as well as increased concentration of the later in eatables, especially vegetables leading to human health hazards. Nitrate itself is not a threat while nitrite is definitely a potential health hazard and that too when found in places at a wrong time. In water bodies, however, nitrate and other forms of N may encourage the growth of algae and subsequently the bacteria leading to exhaustion of molecular oxygen, thereby affecting animal life. Indeed whole ecological balance of water bodies may change due to the so-called "eutrophication".

Consumption of water and vegetables containing excessive amounts of nitrate may lead to the production of nitrite in the stomach and the later becomes particularly dangerous for the babies. Methaemoglobonaemia (blue-baby syndrome) may occur in 1-year old babies taking diet with too much nitrate. Stomach cancer has also been associated with the concentration of nitrate in potable water. Again, it is nitrite that reacts with amines to form N-nitroso compounds, which are reported to cause stomach cancer. Such an illness may result from consumption of vegetables containing high concentrations of nitrate originating from soil or irrigation water.

Nitrification and ecosystem functioning

Nitrification inhibition and consequent accumulation of ammonium would lead to:

1. Increased microbial activities including biological nitrogen fixation, a process that is known to be inhibited more by nitrate -N than ammonium-N.
2. Greater photosynthate partitioning to the rhizosphere thus enriching the soil with organic matter.
3. Enhanced mineralization of native soil N through the so-called priming effect or added nitrogen interaction that is more with ammonium-N than nitrate-N.
4. Increased efficiency of fertilizer N use by plants through decreased losses via denitrification and nitrate leaching.

Microorganisms are known to prefer ammonium over nitrate as a source of N. However, this preference is consequential rather than the reason. This is because of the fact that not all organisms possess nitrate reductase to enable them assimilate nitrate, while almost all of them will be able to assimilate ammonium. In addition, assimilation of nitrate, is more energy intensive as compared to that of ammonium. Hence, for efficient assimilation of nitrate, ample supply of easily oxidizable C will be required. Studies involving the use of glucose as a C source indeed reveal similar assimilation of both ammonium and nitrate by the soil microorganisms. Nevertheless, presence of ammonium leads to an enhancement in microbial activities in terms of respiratory response. In experiments aimed at studying the mineralization of native soil N, ammonium-N is reported to have a significantly higher effect as compared to nitrate-N. This so-called "priming" effect or added nitrogen interaction (ANI) has been found to increase with the amount of applied N. An indirect effect of chemical fertilizers as well as green manures has their positive influence on the mineralization and plant availability of N from the soil organic reserves.

One of the processes reported to be substantially affected by mineral N is the biological nitrogen fixation. This has been unequivocally proved in legumes at each level of the process starting from nodule initiation to actually fixation at the level of nitrogenase. The effect is more intense with nitrate than ammonium leading to studies aimed at successful production of nitrate tolerant legume types. This strategy is particularly important under conditions where legumes are grown together with non-legumes and the later are fertilized with chemical fertilizers. The nature has also endowed the rhizobia with the ability to avoid accumulation of inhibitory levels through dissimilatory nitrate reduction *i.e.*, denitrification.

Plant species vary in their ability to utilize ammonium or nitrate as the sole source of N although a majority of them grow best with a mixture of the two. However, when used as an exclusive source of N, ammonium may cause growth inhibition in many species, particularly in those grown under arable conditions (Marschner, 1999). Under these conditions, nitrification is generally quite rapid and hence deleterious effects of ammonium are avoided. Under saline conditions also, ammonium is reported to increase the sensitivity of plants whereas nitrate has been reported to moderate the negative effects of salinity (Khan *et al.*, 1994). However, the plants are bound to face higher concentrations of ammonium under saline conditions because of the inhibitory effects of salts on the process of nitrification. It would appear therefore that nitrification inhibition would be a blessing for arable plants grown on normal agricultural soils whereas it may be an added problem for those grown on salt-affected lands. Several studies indeed show a positive effect of nitrification inhibitors on plant growth and N use efficiency (Prasad and Power, 1995). The mechanisms involved include i) decrease in losses through denitrification and nitrate leaching, ii) increased microbial activities in the presence of prolonged availability of ammonium and iii) increased rhizpospheric activities etc. Nitrification inhibition may not only decrease in the loss of N through denitrification and nitrate leaching, but also conserve the applied N through enhanced immobilization of the persistently available ammonium-N.

Form of N plays a significant role in affecting root growth, rhizodeposition and the concomitant changes in different rhizospheric microbial functions including root-induced N mineralization. In wheat and maize, root growth may be restricted in ammonium compared to nitrate-fed plants and may be attributed to an increased root respiration, greater allocation of photosynthates to nitrogenous than structural component and increased export of carbon (probably as amino acids) from root to shoot than that occurring under nitrate nutrition. In addition, ammonium nutrition leads to a higher rhizodeposition, thereby enlarging the below-ground sink for photosynthates, most probably at the expense of plant tops thereby reducing the biomass yield. However, increase in rhizodeposition due to increased/sustained availability of ammonium may also prove beneficial to plants in terms of increased microbial activities and especially the mineralization of native soil N. In laboratory experiments, a significant increase in the mineralization of soil N has been observed following addition of easily oxidizable C (Azam and Farooq, 2003).

It has also been suggested above that mineralization of N from soil organic matter is more intense in the presence of ammonium than nitrate. Jenkinson *et al.* (1985) attributed this to "pool substitution" whereby the native N stands proxy for the applied N giving the impression of enhanced mineralization of the later. The fact remains, however, that applied N (especially ammonium) leads to an increase in the availability of soil N. Inhibition of nitrification may therefore lead to a higher mineralization of native soil N thereby augmenting N supplies to plants. In addition, microorganisms responsible for the synthesis of aggregation-adhesion macromolecules may be encouraged by higher availability of carbonaceous materials in the rhizosphere. This will result in better soil structure as well as improved moisture holding capacity of the soil at the root surface. The later may help the plants withstand drought stress at least temporarily. Thus, in spite of the negative effects of ammonium, inhibition of nitrification may still exert beneficial effects on plant growth as mentioned above. The negative effects can be overcome to a significant extent by developing plant types more efficient in using ammonium the so-called "ammoniphilic plants". Plants like rice, sugarcane and kallar grass (*Leptochloa fusca*) etc. growing under high soil moisture conditions can be considered as ammoniphilic plants. Efforts are needed, however, to engineer arable crops (like wheat) for improved tolerance to ammonium while employing nitrification inhibitors.

Another aspect worth consideration is the susceptibility of nitrate to leaching beyond the effective root zone after being converted to calcium nitrate $\{Ca(NO_3)_2\}$ in the presence of ionic calcium (Ca). Hence, in calcareous soils the conditions are quite conducive to this mode of nitrate escape especially following organic amendment that helps in the release of Ca. The leaching is more pronounced in clayey soils at near neutral pH as negative charge on the clays repels nitrate, thereby facilitating the process of leaching. Hence, not only the use efficiency of nitrate will remain low under these conditions but also N economy of the system will be negatively affected.

Management practices to reduce or regulate nitrification

Various approaches have been suggested to improve the use efficiency of nitrogenous fertilizers. These include the improved agronomic practices, use of coatings, chemical additives and the various chemical and physical modifications.

Improved agronomic practices

Split application, placement, foliar applicator fertigation etc. are some of the agronomic techniques by which NUE of nitrogenous fertilizers can be increased (Raun and Johnson, 1999).

Controlled or slow-release fertilizers

By using specific fertilizer formulations to release N in synchrony with plant requirement, it should be possible to provide sufficient N in a single application to satisfy the plant's need, yet maintain very low concentrations of mineral N in the soil throughout the growing season. If this could be done, losses would be small because of the limited amount of N in the substrate.

Several slow-release forms of N (Shaviv and Mikkelsen, 1993) have been suggested. These include:

- *Coated fertilizers*: Soluble urea is coated with an insoluble, slowly permeable but generally biodegradable material to achieve controlled/delayed release of urea-N. Several organic and inorganic coating materials such as sulphur, gypsum, lac, latex, polyolefins, resins, plastics, polyurethanes, rock phosphates etc. have been attempted using three types of coating processes namely, rolling bed, falling curtain and fluidized bed.

- *Complex organic N compounds with relatively less solublity in water than urea*: This group of compounds consisting of urea-formaldehyde complexes (38% N), oxamide (30% N), isobutylidenediurea (IBDU, 30% N), urea-Z (35% N) etc are only slightly soluble in water. The rate of nitrogen release from these compounds depends upon water solubility, microbiological action and chemical hydrolysis.
- *Urea supergranules* (*USG*): The USG consisting of 1-2 discrete urea particles is not so efficient, but its proper deep placement (1 USG for 4 hills at 7-10 cm soil depth with the hole at the placement site closed) makes it efficient.

Many of these fertilizer formulations have been utilized to grow plants in diverse environment. The influence of slow-release forms on levels of soil mineral N, and the recovery of fertilizer N have been assessed for upland crops and lowland rice. The use of these formulations has generally decreased the total loss of fertilizer N.

Use of inhibitors

Fertilizer use efficiency could be greatly increased if the hydrolysis of urea to ammonium by soil urease could be retarded by the use of urease inhibitors, or if nitrate accumulation during the cropping phase could be regulated by nitrification inhibitors.

Urease inhibitors

These reduce the hydrolysis of urea by inhibiting soil urease activity and thus prevent rapid development of high partial pressure of NH_3 and high pH of floodwater in rice fields and eventually reduce NH_3 volatilization losses. A large number of compounds have been tested for their ability to inhibit soil urease but most are ineffective or do not persist in soil. The phosphoroamides, such as phenylphosphorodiamidate (PPD) and N-(n-butyl) thiophosphorictriamide (NBPT), have shown promise for limiting the hydrolysis of urea in laboratory and greenhouse studies when used singly or in combination. Relatively few studies have been done on their ability to reduce NH_3 volatilization and increase grain yield in the field.

Studies using PPD and NBPT as urease inhibitors in flooded rice fields, have shown little reduction in NH_3 loss. The reasons for the failure of PPD in flooded soils seem to be its rapid hydrolysis under the alkaline conditions generated in the floodwater by photosynthetic algae, and its decomposition due to the high temperatures reached in the floodwater. The reasons for the failure of NBPT in flooded soils have not been completely explained, but the results of laboratory studies with non-flooded soils suggest that it must be converted to the oxygen analogue to inhibit urease activity. Studies with another thiophosphorictriamide, thiophosphoryl triamide, showed it too to be a relatively weak inhibitor of urease activity. Appreciable inhibition was achieved only after it had been converted to the oxon analogue. These studies indicate that the thiophosphorictriamides do not inhibit urease activity, but that the phosphorictriamides are potent inhibitors of urease activity.

Field studies in Thailand show that the activity of PPD can be prolonged, and NH_3 loss markedly reduced, by controlling the floodwater pH with the algicide Terbutryn. In addition, a mixture of NBPT and PPD in the presence of Terbutryn was even more effective than PPD alone. It appears that during the time when the PPD was effective, NBPT was being converted to its oxygen analogue. This inhibited urease activity when PPD lost its capacity to do so. The combined urease inhibitor—algicide treatment (Freney *et al.* - http://www.agnet.org/library/article/eb414.html) reduced ammonia loss from 10 to 0.4 kg N-ha^{-1}.

In a laboratory study, cyclohexylphosphorictriamide (CHPT) was found a very effective inhibitor of urease activity and the same was confirmed in a field experiment with flooded rice in Thailand (Freney *et al.* - http://www.agnet.org/library/article/eb414.html). The oxon analogue of NBPT, N-(n-butyl) phosphorictriamide, was compared with CHPT. The two markedly reduced urea hydrolysis, the CHPT being more effective. Its addition maintained the ammoniacal N concentration of the floodwater below 2 g m^{-3} for 11 days, reduced NH_3 loss by -90%, and increased grain yield.

Nitrification inhibitors

The nitrification inhibitors (NIs) decrease the availability of nitrate and consequently its vulnerability to escape mechanisms. Lot of work has been reported on the ways to retard/inhibit the rate of nitrification not only to reduce fertilizer N losses (Prasad and Power, 1995) but also to prolong the persistence of fertilizer N in ammoniacal form (Prasad and Power, 1995; McCarty, 1999). Since ammonia or ammonium-producing compounds are the main source of fertilizer N, maintenance of the applied N in the ammonium form should mean that less N is lost by denitrification. One mechanism of maintaining added N as ammonium is to use a nitrification inhibitor with the fertilizer.

Numerous substances have been tested for their ability to inhibit nitrification (Table 1), and several of these have been patented. Only a limited number of chemicals are available commercially for use in agriculture. These include 2-chloro-6 (trichloromethyl) pyridine (nitrapyrin), sulfathiazole, dicyandiamide, 2-amino-4-chloro-6-methyl pyrimidine, 2-mercapto-benzothiazole, thiourea and 5-ethoxy-3-trichloromethyl-1,2,4-thiadiazole (terrazole). Unfortunately, most of these compounds have limited usefulness. For example, the most commonly used nitrification inhibitor, nitrapyrin, is seldom effective because of absorption on soil colloids, hydrolysis to 6-chloropicolinic acid, and loss by volatilization.

Nitrapyrin: Nitrapyrin [2-Chloro-6-(trichloromethyl)-pyridine, **I**], was developed by Dow Chemical Company. It is marketed under the trade name "N-Serve 24 Nitrogen Stabilizer" (a.i. 240 g L^{-1}) and "N-Serve 24E Nitrogen Stablizer" (a.i. 240 g L^{-1}). The rates of application advised by Dow Chemical Company for band and row placement are 1.125-1.25 L ha^{-1} of N-Serve 24 E for cotton, maize, sugarbeet, sorghum and wheat and 4.50-6.75 L ha^{-1} for potatoes before or after planting or sowing. For broadcasting, the rate of application has to be increased considerably. When granulated fertilizer is used, it can be applied at 0.2-1.0% of the amount of fertilizer N (Slangen and Kerkhoff, 1984). Because of its high vapor pressure, nitrapyrin cannot be granulated with solid-N fertilizer like urea without loss of the inhibitor during processing, storing and handling. Nitrapyrin sometimes shows poor activity due to absorption on soil colloids, hydrolysis to 6-picolinic acid and loss by volatilization.

AM: 2-Amino-4-chloro-6-methylpyrimidine (**II**) is another well-known nitrification inhibitor developed by Toyo Kaotsu Industries Inc. (now Mitsui Toatsu) of Japan. Pure AM is a white crystalline substance (mp, 182°C) and is soluble in water but unlike nitrapyrin, it is relatively insoluble in organic solvents. AM is less volatile and less effective than nitrapyrin. AM is effective when applied @ 5-6 kg ha^{-1} (Prasad and Power, 1995).

Table 1: Some commercial and expensively tested synthetic nitrification inhibitors

Nitrapyrin [2-Chloro-6-(trichloromethyl)-pyridine]
2-Amino-4-chloro-6-methylpyrimidine (AM)
Dicyandiamide (DCD)
Etridiazole (5-Ethoxy-3-trichloromethyl-l,2,4-thiadiazole)
N-(2,5-Dichlorophenyl succinamic acid) (DCS)
Potassium azide (KN_3)
3-Amino-l,2,4 triazole(ATC)
Thiourea (TU)
MBT (2-Mercaptobenzothiazole)
2-Ethynyl pyridine
MPC (3-Methyl-pyrazole-l-carboxamide)
ST (2-sulfanilamidothiazole)
Carbon disulphide
3-Mercapto-1,2,4-triazole
Sodium diethylthiocarbamate
Acetylene
Gaseous hydrocarbon e.g.-methane, ethane, ethylene
Ammoniu n thiosulphate
Thiophosphoryl triamide
4-Mesyl benzotrichloride
4-Nitrobenzotrichloride
Guanyl thiourea
2,4-Diamino-6-trichloromethyl triazine
Potassium trithiocarbonate
Sodium thiocarbonate
2-Amino-4-methyl-6-trichloromethyltriazine (MAST)
2-Benzothiazole sulfane morpholine
3,4-Dimethyl pyrazole phosphate

Etridiazole: 5-Ethoxy-3-trichloromethyl-l,2,4-thiadiazole (Terrazole, Etridiazole, Dwell, **III**) is an effective nitrification inhibitor developed by Olin Corporation, Baltimore, USA. This product is available as a wettable powder or technical grade liquid with 35% and 95% a.i., respectively. As a coating on ammonium sulphate and urea, terrazole 95% a.i. is used upto 1.5% by weight. The recommended rates of compound for crops like potatoes, sugarbeet, lettuce and onions are 0.6-1 kg ha^{-1} (Slangen and Kerkhoff, 1984). Besides this compound, some other thiodiazoles are also known to inhibit ammonia oxidation. Among them 3,4-dichloro-l,2,5-thiadiazole is noteworthy.

Dicyandiamide (DCD): It (IV) has been developed both as a slow release nitrogen source as well as nitrification inhibitor (Slangen and Kerkhoff, 1984). In Japan, it is added to mixed fertilizer and a product urea-form plus containing 10% by weight of DCD is produced. A fertilizer containing urea and DCD in a 4:1 ratio is commercially available in West Germany. DCD is toxic to plants but the effects differ with plant species. This compound is effective over a period of 1-3 months. When applied @ 10-15% of applied nitrogen, it remains active for a period of 2 months. Cost of production of DCD is lower than the corresponding cost of nitrapyrin and etridiazole and DCD has the advantage that it completely decomposes in soil to CO_2 and NH_4^+ over several weeks and thereby acts as a high analysis (66.7% N) slow release N fertilizer. Compounds containing ammonium sulphate and dicyandiamide are available in granulated and coated form from Suddeutschekalksticksoffwerke AG, Trostberg, Germany and Sisco Corporation, Japan and recommended as slow release fertilizers. Increase in NUE of urea by DCD in field studies on different crops has been reported by various authors (Slangen and Kerkhoff, 1984).

CMP: [l-Carbamoyl-3-methylpyrazole, V] has been found to be an effective nitrification inhibitor (McCarty, 1999). Under flooded conditions, CMP affected almost total inhibition of nitrification and prevented build up of nitrite and nitrate in floodwater. In pure culture studies, CMP inhibited the growth of *Nitrosomonas* at concentrations as low as 1 ppm.

ST: *A* group of thiazoles such as sulfathiazoles and especially 2-sulfanilamidothiazole (ST, VI) were introduced as nitrification inhibitors by Mitsui Toatsu Chemicals Inc., Japan. ST is more stable than AM and can be formulated with both acidic and basic fertilizers. It is apparently more volatile than AM. ST is commercially used in Japan on a limited scale.

ATC: Many triazoles particularly 1,2,4-triazoles are reported to have nitrification inhibition activity. Among these, 4-amino-1,2,4-triazole (ATC, VII) is the most potent inhibitor but not as effective as nitrapyrin. ATC was produced from formic acid and hydrazine by Ishihara Industries, Japan. It completely checks nitrification of urea for four weeks at a concentration of 5% by weight of urea.

Pyriilines: Among the various pyridines tested, 2-ethynylpyridine (VIIT) and nitrapyrin (I) are the most potent inhibitors of nitrification but other compounds also possess this activity. Among those, 2-chloropyridine, 2,6-dichloropyridine, 6-chloro-2-picoline significantly inhibited ammonia oxidation in soil whereas compounds containing carboxylic group (6-chloro-picolinic acid) had little effect (McCarty and Bremner, 1989).

3,4-Dimethylpyrazole phosphate (DMPP): It (general structure, IX) is a new nitrification inhibitor with highly favourable toxicological and ecotoxicological properties and shows several distinct advantages compared to the currently used nitrification inhibitors. Application rates of 0.5-1.5 kg ha^{-1} are sufficient to achieve optimal nitrification inhibition. It can significantly reduce nitrate leaching, without being liable to leaching itself (Zerulla *et al.*, 2001).

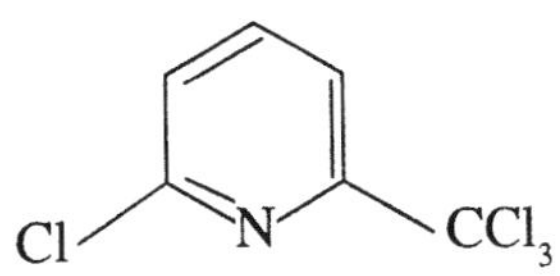

(I)

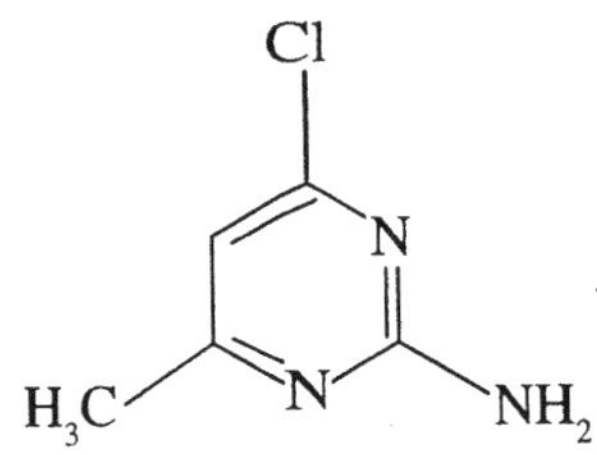

(II)

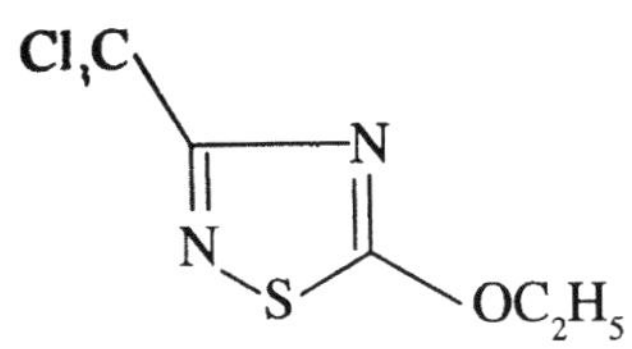

(III)

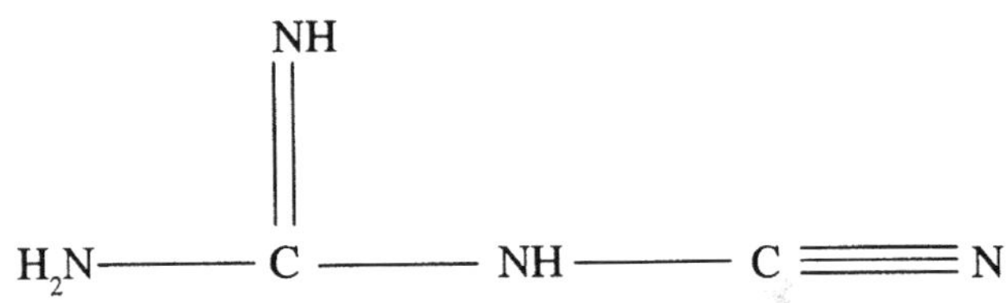

(IV)

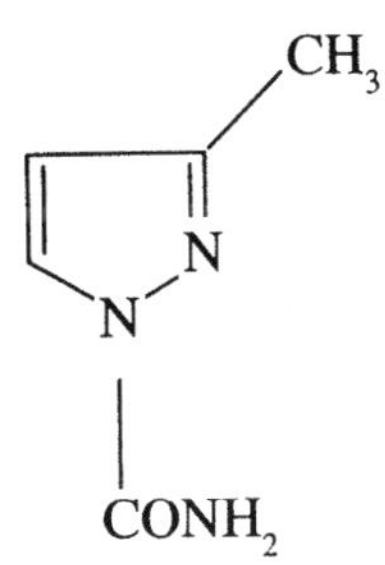

(V)

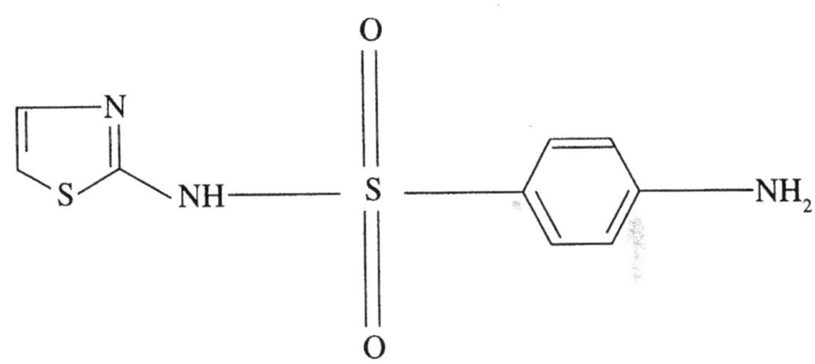

(VI)

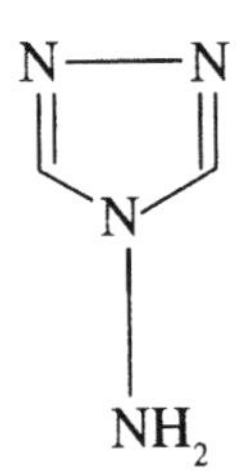

(VII)

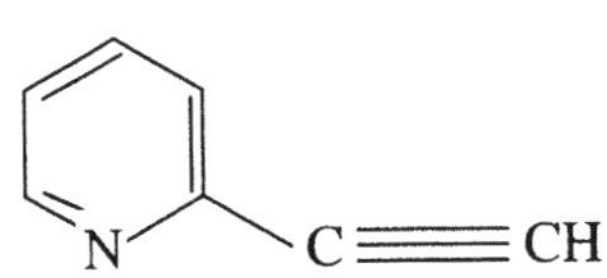

(VIII)

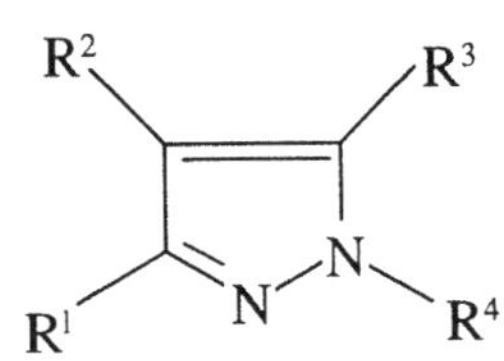

(IX)

Some commercial nitrification inhibitors

Furan derivatives: Sahrawat and Mukerjee (1977) after observing the efffect of furan ring on nitrification inhibition screened some furano compounds like furfural (X) and furfural alcohol (XI) for possible effect on nitrification. Twenty to 30% concentrations of these compounds matched 5% karanjin (the major furanoflavonoid constituent of *Pongamia glabra)* in nitrification inhibition during 45-60 days. The inhibition of nitrification decreased after 45 days while karanjin remained effective even after this period. Furfural alcohol was better inhibitor of nitrification of ammonium sulphate than that of urea.

Kuzvinzwa (1984) tested derivatives of furfural along with a natural furanocoumarin psoralen, for nitrification inhibition in laboratory incubation studies. 5-Nitrofurfural oxime (XII), furfural oxime (XIII) and furfural semicarbazone (XIV) were the most effective followed by 5-nitrofurfural semicarbazone (XV), 5-nitro-3-chloro-2-furanilide (XVI) and psoralen (XVII). Only 5-nitrofurfural oxime approached nitrapyrin in effectiveness. The nitro derivatives tended to become general bactericides and became effective against *Nitrobacter* species also, thereby causing accumulation of nitrite nitrogen. 3-Chlorofurananilide and furfural oxime caused very little accumulation of nitrite even at the highest concentration (15%, N-basis).

Datta *et al.* (2001) reported three series of furfural derivatives, namely N-O-furfural oxirne ethers (XVIII), furfural Schiiff bases (XIX), and furfural chalcones (XX), as possible nitrification inhibitors in laboratory incubation study. Furfural oxime ethers and Schiff bases showed potential activity, but furfural chalcones were only mildly active. N-O-Ethyl furfural oxime among the oxime ethers, and furfurylidine-4-chloroaniline among Schiff bases, performed the best. These two compounds showed more than 50% nitrification on the 45th day at 5% dose as compared to 73% by nitrapyrin.

Activity of the ethers decreased with increase in N-O-alkyl chain length and introduction of chlorine in phenyl ring of furfurylidine anilines increased the activity of Schiff base.

Acetylenic compounds: Acetylene was first found to inhibit ammonia oxidation in pure cultures of *Nitrosomonas europea* and then established as a potent inhibitor of nitrification (McCarty, 1999). However, because it is a gas, there are problems in introducing it into the soil in field, and sustaining its availability during the growing period at a concentration required to limit nitrification. The problem has been overcome by the use of calcium carbide coated with layers of wax and shellac to provide a slow-release source of acetylene. Addition of wax-coated calcium carbide to the fertilized soil has reduced nitrification and increased yield, or recovery of N, in irrigated wheat, maize, cotton and flooded rice (Banerjee *et al.,* 1990).

Another way of overcoming the problem of applying gaseous acetylene is to use substituted acetylenes such as 2-ethynylpyridine or phenylacetylene, which are liquids at ambient temperatures. These two compounds have proved to be effective inhibitors in laboratory studies. The use of 2-ethynylpyridine in irrigated cotton has resulted in greatly increased recovery of applied N (Freney *et al.* - http//www.agnet.org/library/article/eb414.html).

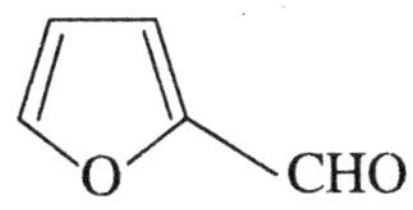

(X)

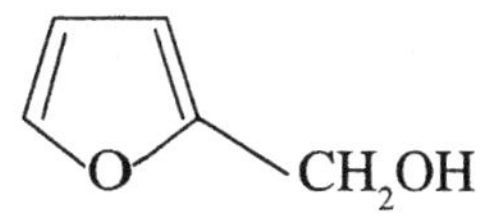

(XI)

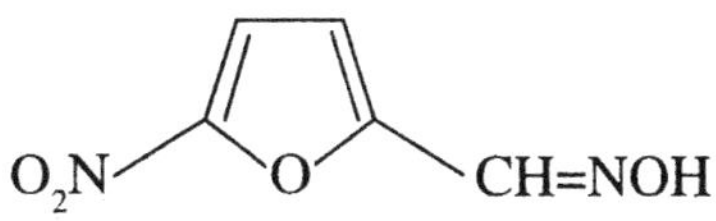

(XII)

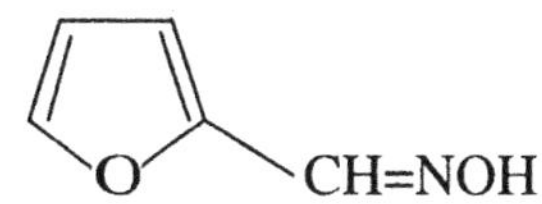

(XIII)

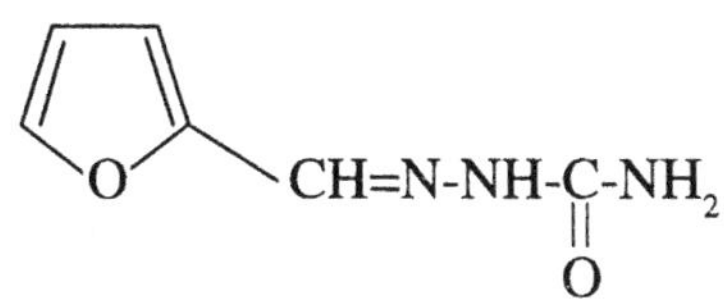

(XIV)

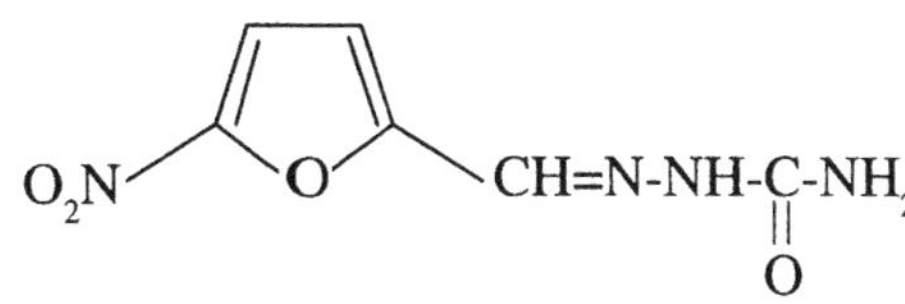

(XV)

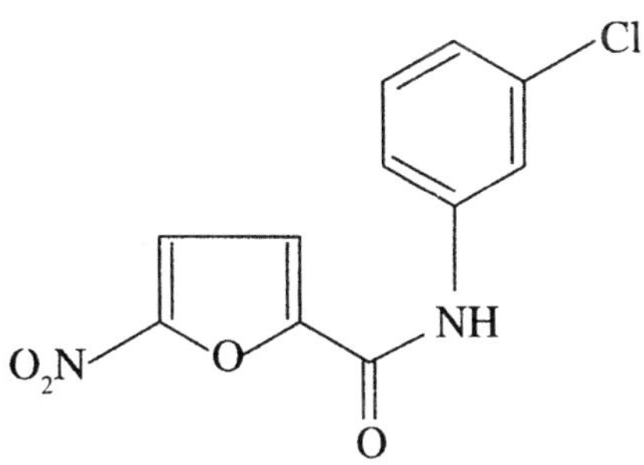

(XVI)

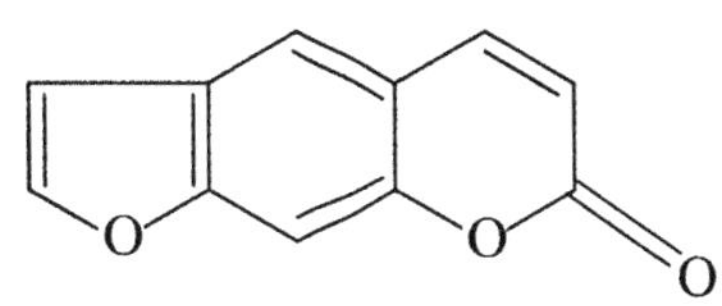

(XVII)

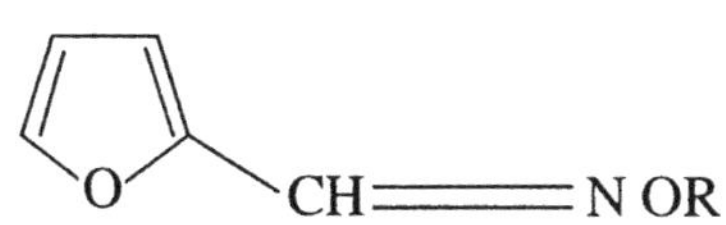

R=C_2H_5/$CH(CH_3)_3$/C_4H_9/ C_6H_{11}/CH_2-CH=CH_2

(XVIII)

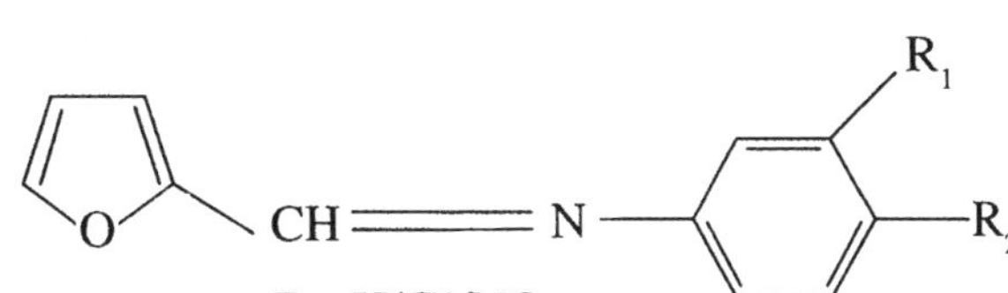

R_1=H/Cl/NO_2
R_2=H/CL

(XIX)

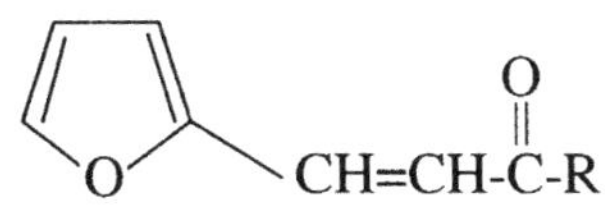

R=CH_3/C_6H_5/4-OH-C_6H_5

(XX)

Furfural derivatives with nitrification inhibitory property

Sulphur compounds. A broad range of S containing compounds including thiosulphates, thiocarbamates, xanthates, S-containing amino acids, and several pesticides and fungicides inhibit nitrification. Specific compounds include: carbon disulphide (CS_2), thiourea, allyl thiourea, gaunyl thiourea, 2-mercapto benzothiazole, 3-mercapto-1,2,4-triazole, thioacetamide, sodium diethyldithiocarbamate, sodium thiocarbonate, thiosemicarbazide, thiocarbohydrazide, diphenyl thiocarbazone, dithiocarbamate, s-ethyl dipropyl thiocarbamate, ethylene-*bis*-dithiocarbamate, and N-methyl dithiocarbamate.

Heterocyclic compounds: Several strong inhibitors of ammonia oxidation in soil can be classified by their heterocyclic ring structures. This classification includes some of the more potent inhibitors of nitrification in soil, namely nitrapyrin, etridiazole, 2-ethynyl pyridine, 4-amino-l,2,4-triazole, 3-methylpyrazole-1-carboxamide and recently reported furan derivatives (Datta *et al.*, 2001), 1,3,4-oxadiazoles (Kumar *et al.*, 2005), 3,4-dimethyl pyrazole phosphate (Zerulla *et al.*, 2001). The heterocycles involved in general are: furan, pyrazole, pyridine, pyridazine, benzotriazole, 1,2,4-triazole, thiadiazoles, 1,3,5-triazines, s-tetrazines. Several of the heterocyclic N compounds found to inhibit ammonia oxidation are structurally similar in that they contain chloro (Cl) and/or trichloro methyl (CC13) groups substituted on carbon atom(s) adjacent to a ring N [*e.g.*, nitrpyrin {2-chloro-5-(trichloromethyl) pyridine}, etridiazole (5-ethoxy-3-trichloromethyl-l,2,4-thiadiazole), 2-chloropyridine, 2,6-dichloropyridine, 6-chloro-2-picoline, and 3,4-dichloro-1,3,4-thiadiazole].

Inhibitors of natural origin: Synthetic nitrification inhibitors, though expensive can efficiently inhibit nitrification. Certain allelochemicals released by plants are also reported to have an inhibitory effect. Rice (1984) postulated that because inhibition of nitrification results in conservation of both energy and nitrogen, vegetation in late succession or climax ecosystems contains plants that release allelochemicals that inhibit nitrification in soil. Some natural products from neem (*Azadirachta indica,* A. juss), karanja (*Pongamia glabra,* Vent.), mint (*Mentha spicata, M. arvensis* L.) and mahua (*Madhuca longifolia,* L.) are reported to inhibit the activity of nitrifiers. Among them, neem based products like Nimin, Neemex and Neem Gold-A are commercially available in the Indian market.

Nonedible oil seeds like neem (*Azadirachta indica* A. Juss.), karanj (*Pongamia glabra* Vent.), mahua (*Madhuca indica, M. latifolia*) have been extensively studied for nitrification inhibition properties. Oil cakes in general and nonedible cakes in particular have been known to possess certain minor non fatty biologically active constituents which make them unsuitable for human consumption. These oilseeds have been traditionally used as slow nutrient release manures or in admixture with manures to regulate the nutrient release.

Neem: The utility of neem cake in improving the nitrogen use efficiency of prilled urea in different crops has been exhaustively demonstrated by several workers (Prasad *et al.,* 1993). Due to the poor shelf life of neem cake and the low N content of cake coated urea, its industrial production did not merit attention (Prasad *et al.,* 2002). Neem oil coated urea (NOCU) on the other hand was found to be more suited (Saxena *et al.,* 1999). Its efficacy has been demonstrated at two-fertilizer plants in India *viz.* KRIBHCO, Hazira and

Sri Ram Fertilizers and Chemicals, Kota. National Fertilizer Ltd., Panipat have independently produced neem oil coated urea.

Karanja: A number of physiologically active furano-flavonoids are found in seeds, bark and leaves of the tree *Pongamia glabra* Vent. (Sahrawat *et al.,* 1974). The hot ethanol extract of the defatted seeds (applied @ 20 and 30% of applied N) had maximum nitrification inhibitory activity followed by bark extract (prepared with 40: 60, v/v, mixture of petroleum ether and acetone), and leaves had negligible effect. The alcohol extract of seeds was effective in retarding nitrification for 60 days when applied at 20% of the fertilizer nitrogen dose. The percentage inhibition of nitrification of urea in soil was 47-55% even after 45 days of application of seed extract.

Karanjin, present in karanj seeds, is a potent inhibitor of nitrification. The inhibition of nitrification of urea or ammonium sulphate was around 43% after 8 weeks of incubation with it at dose of 5% of applied fertilizer nitrogen. Comparative evaluation of karanjin and three commercial inhibitors (nitrapyrin, AM and DCD) when applied in a sandy loam soil at 5 mg kg^{-1} of soil reduced the nitrification rate as:

$$\text{Nitrapyrin} > \text{Karanjin} > \text{AM} > \text{DCD}$$

The furan ring of karanjin was responsible for nitrification inhibition property. This information lead to the study of several furan derivatives as nitrification inhibitors (described in synthetic compounds).

Mahua: Seed cake and extracts of *Bassica latifolia* Roxb., *Madhuca indica* G mel., *Madhuca latifolia* L. are known to possess nitrification inhibition property. The seed cake and extracts contain saponins, responsible for nitrification inhibition. In an incubation study employing clay loam soil, the inhibitory effect of mahua cake extract persisted only for 20 days (Slangen and Kerkhoff, 1984).

Miscellaneous: Many other plants and plant products from *Citrullus colosynthis* Schrad. Sal *(Shorea robusta), Eucalyptus globosus, Ricinus communis. Acacia catechu, Calotropis gigantean, Onosma hispidium, Mentha arvensis, Mentha spicata, Artemisia annua, Chrysanthemum cinerarifolium, Tagetes erecta, Catharanthus roseus, Ricinus communis L.,* turmeric powder, tea waste and cashew shell powder have been reported to show varying degree of nitrification inhibition.

Conclusion

Several nitrification inhibitors have been reported in literature from time to time. Most of these are not still fully satisfactory due to one or more of the following disadvantages:

- Complicated synthesis and related preparation steps
- High volatility, leading to low persistence and high losses into the atmosphere unless introduced in soil using technically complicated processes (*e.g.* by probe)
- High toxicity or ecotoxicity
- Low stability against hydrolysis, reducing the duration of action in soil and the shelf life
- High application rates

- Requirement of further modification of the active molecule
- Addition of costly formulants
- High cost

Therefore, an ideal nitrification inhibitor is still elusive. It needs to be simple, safe, efficient, persistent, specific and economical in use. It implies that the nitrification inhibitor should have specificity to nitrifying bacteria responsible for conversion of ammonium to nitrite. Inhibition of *Nitrobacter* is not desirable as it leads to accumulation of nitrite. The inhibitor should be non-toxic to other soil organisms, fish, mammals, crops and be safe to the environment. It should be effective throughout the nitrogen-soil interaction zone and be sufficiently persistent in action so that nitrification is inhibited for an adequate period of time. Furthermore, it should be a low cost additive to the fertilizer.

A major consideration during the selection of nitrification inhibitors (NI) is their high effectiveness at the lowest possible application rate with a minimum of undesirable side effects. The availability of an inhibitor at effective concentration is essential. This can be achieved by coating fertilizer granules with the inhibitor or by incorporating it into granules (Zerulla *et al.,* 2001). The aim of both the approaches is to ensure an intimate and uniform interaction of the substrate with the inhibitor. The application of an effective concentration of a NI to soil, together with N fertilizer, is a difficult task since it involves different crops, forms and rates of N application. It leads to different concentrations of NI reaching the nitrifiers, particularly if N is applied as granules.

While application of chemical fertilizers to agricultural crops has resulted in tremendous increase in yield, problems arising due to escape to the environment of different nitrogen species, especially N_2O, nitrite and nitrate, have raised serious economic and environmental concerns. Of the different processes responsible for these, nitrification and denitrification are of prime importance. Hence, efforts have to be made to regulate the process of nitrification (major source of different N species) as a means to enhancing the use efficiency of N, decreasing environmental/economic concerns and optimizing the functioning of agro-ecosystems. Use of nitrification inhibitors has been helpful in mitigating the negative effects of fertilizer application. However, continued efforts need to be made for finding more efficient and environment friendly products to suit the everchanging agro-climatic conditions.

References

Abbasi, M.K. and Adams, W.A. 2000. Estimation of simultaneous nitrification and denitrification in grassland soil associated with urea-N and ^{15}N and nitrification inhibitor. *Biol. Fertil. Soils,* **31**: 38-44.

Azam, F., Mueller, C., Weiske, A, Benckiser, G. and Ottow, J.C.G. 2002. Nitrification and denitrification as sources of atmospheric N^2O - role of oxidizable C and applied N. *Biol. Fertil. Soils,* **35**: 54-61.

Azanm F, abd Farooq, S. 2003. Nitrification inhibition in soil and ecosystem functioning—An overview. *Pak. J. Biol. Sci.,* **6**(6): 528-535.

Banerjee, N.K., Mosier, A.R., Uppal, K.S. and Goswami, N.N. 1990. Use of encapsulated calcium carbide to reduce denitrification losses from urea-fertilized flodded rice. *Mitteihungen der Deutschen Bodenkundlichen Gesellschaft,* **60**: 245-248.

Bedard, C. and Knowles, R. 1989. Physiology, biochemistry, and specific inhibitors of CH^4, NH_4^+, and CO oxidation by methanotrophs and nitrifiers. *Microbial Reviews,* **53**: 68-84.

Bothe, H., Jost, G., Schloter, M., Ward, B.B. and Witzel, K.P. 2000. Molecular analysis of ammonia oxidation and denitrification in natural environments. *FEMS Microbiology Reviews,* **24**: 673-690.

Datta, A., Walia, S. and Parmar, B.S. 2001. Some furfural derivatives as nitrification inhibitors. *J. Agric. Food Chem.,* **49**: 4726-4731.

Duddleston, K.N., Bottomley, P.J., Porter, A.J. and Arp, D.J. 2000. New insights into methyl bromide cooxidation by *Nitrosomonas europaea* obtained by experimenting with moderately low density cell suspensions. *Applied Environ. Microbiol.,* **66**: 2726-2731.

Freney, J.R., Peoples, M.B. and Mosier, A.R. Efficient use of fertilizer nitrogen by crops. http://www.agnet.org/library/article/eb414.html.

Holmes, A.J., Costello, A., Lindstrom, M.E. and Murrell, J.C. 1995. Evidence that particulate methane monooxygenase and ammonia monooxygenase may be evolutionary related. *FEMS Microbiology Letters,* **132**: 203-208.

Jenkinson, D.S, Fox, R.H. and Rayner, J.H. 1985. Interactions between fertilizer nitrogen and soil nitrogen—the so-called "priming" effect. *J. Soil Sci.,* **36**: 425-444.

Khan, M.G., Silberbush, M. and Lips, S.H. 1994. Physiological studies on salinity and nitrogen interaction in alfalfa. I. Biomass production and root development. *J. Plant Nutr.,* **17**: 657-668.

Kumar, R., Anupama and Parmar, B.S. 2005. Process for the preparation of 5-substituted-1,3,4-oxadiazole-2-thiols as new urease and nitrification inhibitors. Patent submitted to ICAR.

Kuzvinzwa, S.M., Devakumar, C. and Mukherjee, S.K. 1984. Evaluation of furano compounds as nitrification inhibitors. *Bull. Indian Soc. Soil Sci.,* **13**: 165-172.

Marschner, H. 1999. *Mineral nutrition of higher plants.* 2nd Edn. Academic Press, London.

McCarty, G.W. 1999. Modes of nitrification inhibitors. *Biol. Fertil. Soils,* **29**: 1-9.

McCarty, G.W. and Bremmer, J.M. 1989. Inhibition of nitrification in soil by heterocyclic nitrogen compounds. *Biol. Fertil. Soils,* **8**: 204-211.

Prasad, R. 1998. Fertilizer urea, food security, health and the environment. *Curr. Sci.,* **75**: 677-683.

Prasad, R. and Power, J.F. 1995. Nitrification inhibitors for agriculture, health and the environment. *Adv. Agron.,* **54**: 233-281.

Prasad, R., Sharma, S.N., Singh, S., Devakumar, C. and Saxena, V.S. 2002. Neem coating of urea for the environment and agriculture. *Fertil. News,* **47**(Part 5): 63-67.

Prasad, R., Devakumar, C. and Sivay, Y.S. 1993. Significance in increasing fertilizer nitrogen efficiency. In: *Neem Research and Development* (Eds. N.S. Randhawa and B.S. Parmar), pp. 97-108. SPS Publication No. 3, Society of Pesticide Science, India, New Delhi.

Purkhold, U., Pommereninger-Röser, A., Juretschko, S., Schmid, M., Koops, H.P. and Wagner, M. 2000. Phylogeny of all recognized species of ammonia oxidizers based on comparative 16S rRNA and *amoA* sequence analysis: implications for molecular diversity surveys. *Applied and Environmental Microbiology,* **66**: 5368-5382.

Raun, W.R. and Johnson, G.V. 1999. Improving nitrogen use efficiency for cereal production. *Agronomy J.,* **91**: 357-363.

Rice, E.L. 1984. *Allelopathy,* 2nd Ed., Academic Press, New York.

Sahrawat, K.L., Parmar, B.S. and Mukherjee, S.K. 1974. A note on the nitrification inhibitors in the seeds, bark and leaves of *Pongamia glabra* Vent. *Indian J. Agric. Sci.,* **44**: 415-418.

Sahrawat, K.L. and Mukherjee, S.K. 1977. Nitrification inhibitors: I. Studies with furano compounds. *Plant Soil.,* **47**: 687-691.

Saxena, V.S., Devakumar, C. and Prasad, R. 1999. Pusaneem-ME coated urea. *Indian Patent* Application No. 223/Del/99 dated 5th Feb. 1999.

Shaviv, A. and Mikkelsen, R.L. 1993. Controlled release fertilizers to increase efficiency of nutrient use and minimize environmental degradation: a review. *Fert. Res.* **35**: 1-12.

Slangen, J.H.G. and Kerkhoff, P. 1984. Nitrification inhibitors in agriculture and horticulture: a literature review. *Fert. Res.,* **5**: 1-76.

Williams, P.H., Jarvis, S.C. and Dixon, E. 1998. Emission of nitric oxide and nitrous oxide from soil under field and laboratory conditions. *Soil Biol. Biochem.* **30**: 1885-1893.

Zerulla, W., Barth, T., Dressel, J., Erhardt, K., Horchler von Loquenghien, K., Pasda, G., Rädle, M. and Wissemeier, A.H. 2001. 3,4-Dimethylphyrazole phophate: a new nitrification inhibitor for agriculture and horticulture. *Biol. Fertil. Soils.,* **34**: 79-84.

Soil Microbiology and Soil Biotechnology : Vol. 1
Editor : S. Kannaiyan
Associated Publishing Co., New Delhi, 2010, pp. 63-75

5

Denitrification in Soils, Constructed Wetlands and Waste Water Treatment Plants

Benckiser

Denitrification and the Nitrogen Cycle

Soils are composed of many physically and (bio)chemically diverse subenvironments, which provide niches for billions of prokaryotic bacteria (Eubacteria and Archaea), and eukaryots as fungi, protozoa or nematodes (Benckiser and Schnell, 2006). Heavy rain falls or irrigation events change the soil oxgen contents dramatically. Anaerobic respiration processes begin to play a dominant role in organic matter and nitrogen cycling. Nitrogen is introduced into the N-cycle by biological and technical N_2-fixation (Fig. 1). Nitrogen-fixing bacteria depend highly on a good carbon availability when they are carrying out the cleavage of the threefold bonded $N \equiv N$ from the atmospheric pol consisting of 78% N_2 (v/v), about 21% O_2 (v/v), 0,03% CO_2 (v/v) and traces (ppb - ppm) of NO and N_2O, because this process needs a lot of energy. Symbiontic cooperation is most efficient in providing high energy inputs. N_2 transformed into NH_4^+ or NH_2-containing organic compounds fills after desamination of NH_2-groups by a broad spectrum of microorganisms the soil NH_4^+-pool, the crossroad of major N processes (Fig. 1). Autotrophic bacteria oxidize the ammonium not incorporated into proteinous compounds to NO_3^- for conserving energy. Heterotrophic nitrifiers oxidize organically bound NH_2 to NO_2^-. At it was recently found can autotrophic nitrifying bacteria also denitrify (nitrifier-denitrification; Liz *et al.*, 2006). NH_4^+ serves as electron donator and NO_3^- as electron acceptor. The exergonic oxidation of NH_4^+ to NO_3^- and the exergonic anaerobic reduction of NO_3^- with NH_4^+ as electron acceptor to gaseous N species combined in one bacterial cell improve the survival considerably. Interesting in this respect is that nitrifierdenitrification contributes more ATP to the cell energy budget than the aerobic nitrification process (Benckiser, 1997; Liz *et al.*, 2006). Denitrification either autotrophic or heterotrophic aims to conserve energy by generating electrochemical gradients across the cytoplasmic membrane through e^- transfers from NH_4^+ or organic compounds on the redox couples NO_3^-/NO_2^-, NO_2^-/N_2O, NO/N_2O and N_2O/N_2 (standard redox potentials, E_0', +432, +759, +1177 or +1352 mV, respectively, (Thauer *et al.*, 1977). The denitrification process ends with the release of gaseous N-products

($NO_2\uparrow, NO\uparrow, N_2O\uparrow, N_2\uparrow$). Most abundant end product is N_2. $NO\uparrow$ and $N_2O\uparrow$ are released in minor concentratiosn (ppb - ppm range). Dissimilatory nitrate ammonification and the not electrogenic assimilatory ammonification, other types of nitrate reduction, end with $NH_3\uparrow$. The not electrogenic assimilatory ammonification pathway uses nitrite and NO as electron sinks for detoxifying $NO_2^{\bullet}$ and $NO^{\bullet}$ radicals. In the energy conserving dissimilatory reduction of nitrate to N_2 or NH_3 several terminal, seemingly modular organized oxidoreductases are involved and within a bacterial cell there is apparently no option to denitrify and to ammonify nitrite simultaneously (Zumft, 1997; Benckiser *et al.*, 2005). N_2O released by denitrification is as N_2 a relatively inert gas that accepts electrons only in the presence of an activating agent, usually a transition metal. Emitted into the atmosphere N_2O turns out as an important greenhouse gas. N_2O, a gas relatively inert as N_2, reaches the troposphere where it converts energized through the ultraviolet light spectrum of the sun to NO. NO reacts with ozone to nitrite and returns as HNO_2 (acid rain) to the earth.

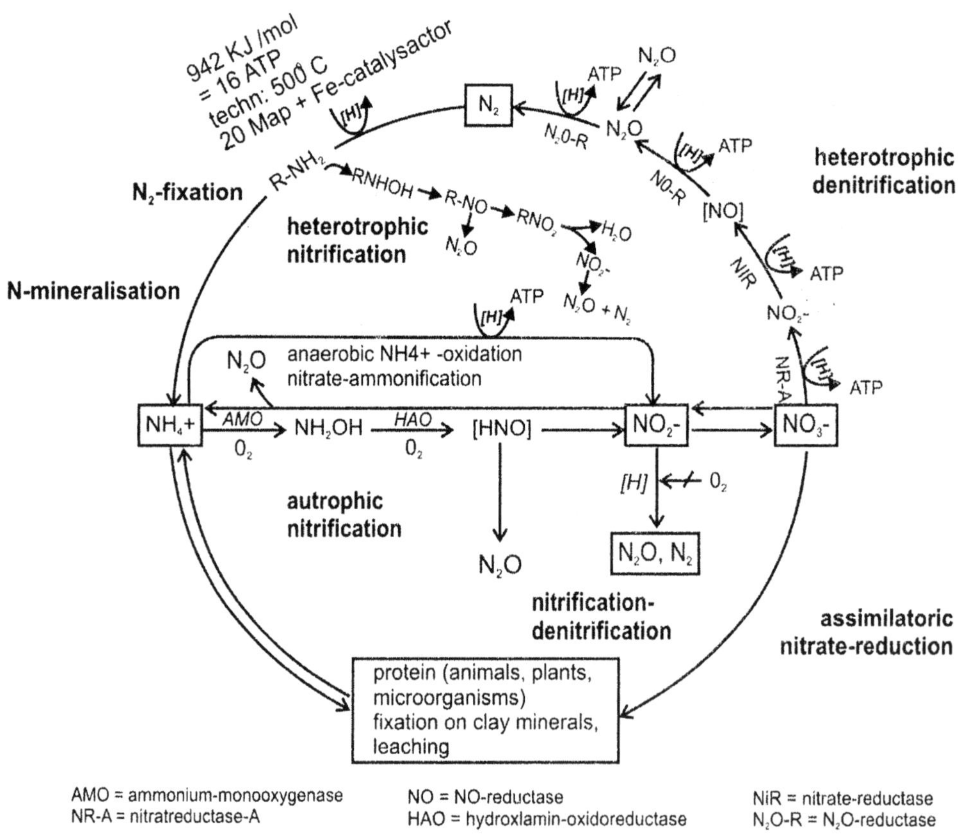

Fig 1: The Nitrogen Cycle

Bacteria can use the redox pair 0.5 O_2/H_2O (aerobic respiration) and besides N-oxides $Mn^{4+}/Mn^{2+}, Fe^{3+}/Fe^{2+}, SO_4^{2-}/H_2S, CO_2/CH_4$ (anaerobic respiration). Internal organic intermediates serve as electron acceptors during fermentation (Thauer *et al.*, 1977).

Only prokaryotic microorganisms have these broad spectrum of energy conserving pathways available which are sometimes interwoven as the newly discovered nitrate reduction pathway mediated by Fe(III)-reducing and Fe(II)-oxidizing bacteria reveals. This process crisscrosses during the redox cycling denitrification with Fe^{3+}-reduction and releases a mixture of N_2O, N_2 and NH_4^+ from transient accumulating NO_2^- (Weber *et al.*, 2006). With the mentioned aerobic and anaerobic respiration and fermentation strategies microorganisms can use nearly all known aliphatic and aromatic carbon sources, reduced S compounds or molecular hydrogen as electron donators and grow even under extreme conditions as hot or arctic temperatures, high salt concentrations, anaerobiosis, or pH values between 1 to 12. The thermodynamic and biochemical details behind denitrification are found by Thauer *et al.* (1977), Zumft (1997); Benckiser (1997), and in microbial text books or numerous original articles.

Table 1: Microbial genera suggested to include denitrifiers (Burth *et al.*, 1982; Albrecht *et al.*, 1997; Zumft, 1997; Philippot 2005; Shapleigh, 2005)

Archaea:	*Haloarcula, Halobacterium, Haloferax, Ferroglobus, Pyrobaculum*
Bacteria:	
Gram-negative:	*Aquiflex, Flexibacter, Empedobacter, Flavobacterium, Sphingobacterium, Synechcystis sp.* PCC 6803.
α-subdivision:	*Agrobacterium, Aquaspirillum, Azospirillum, Blastobacter, Bradyrhizobium, Gluconobacter, Hyphomicrobium, Magnetospirillum, Nitrobacter, Paracoccus, Rhizobium, Rhodobacter, Rhodoplanes, Rhodopseudomonas, Roseobacter, Sinorhizobium, Thiobacillus.*
β-subdivision:	*Achromobacter, Acidovorax, Alcaligenes, Azoarcus, Brachymonas, Burkholderia, Chromobacterium, Comamonas, Eikenella, Hydrogenophaga, Janthinobacterium, Kingella, Microvirgula, Neisseria, Nitrosomonas, Ochrabacterium, Oligella, Ralstonia, Rubrivivax, Thauera, Thermothrix, Thiobacillus, Vogesella, Zoogloea.*
γ-subdivision:	*Acinetobacter, Alteromonas, Azomonas, Beggiatoa, Deleya, Halomonas, Marinobacter, Moraxella, Shewanella, Thioploca, Thiomargarita, Xanthomonas, Pseudoalteromonas, Rugamonas, Pseudomonas* (rRNA group I: *P. aeruginosa, P. alcaligenes, P. aureofaciens, P. balearica, P. chlororaphis, P. fluorescens* bv. B, C, F, *P. mendocina, P. mucidolens, P. pseudoalcaligenes, P. stutzeri;* rRNA group II: *P. nitroreducens, P. caryophylli, P. mallei, P. pickettii, P. plantarii, P. pseudomallei;* Species awaiting grouping or incertae sedis: *P. azotoformans, P. butanovora, P. carboxydoflavac, P. carboxydohydrogenac, P. gazotrophac, P. indigofera, P. lemoignei, P. marginalis, P. nautica*).
δ- subdivision:	none
ξ-subdivision:	*Wolinella, Campylobacter, Thiomicrospira*
Gram-positive:	*Bacillus licheniformis, B. stearothermophilus, B. halodenitrificans, B. firmus, B. cereus, B. azotoformans, B. macerans, Corynebacterium, Frankia, Dactylosporangium, Dermatophilus, Gemella, Jonesia* (formerly *Listeria*), *Kineosporia, Micromonospora, Microtetraspora, Norcardia, Pilimelia, Proprionibacterium, Saccharomonospora, Saccharothrix, Spirillosopora, Saccharothrix, Spirillospora, Streptomyces, Streptosporangium.*
Fungi:	*Fusarium, Giberella, Trichoderma, Cylindrocarpon, Chaetomium, Penicillium, Aspergillus* (fungi imperfecti); *Cylindrocarpon, Talaromyces* (filamentous fungi); *Hansenula* (yeasts).

It is estimated that about flow annually through the global nitrogen cycle. Heterotrophic denitrifiers, a group rich in species, while autotrophic nitrifiers, a group of relatively few species (Stuven and Bock, 2001), assimilatoric and dissimilatoric nitrate ammonifiers, and a few fungi (Table 1) return nitrogen not fixed in organic matter or in clay minerals into the atmosphere.

Heterotrophic denitrifiers have the capability to use anaerobically in the presence of nitrate a broad spectrum of electron donators comprising sugars, alcohols, aldehydes, aromatic ring structures as toluene, xylene, phenols, cresols, phthalate, cyclohexanol, benzoate, or even aromatic compounds as recalcitrant as chlorobenzene as electron donators (Retering *et al.*, 2006; Benckiser and Schnell, 2006) and are therefore increasingly biotechnologically used in municipal and industrial waste water treatment plants. Because organic hydrocarbons, alkanes and aromatic compounds are more efficiently degradated under O_2 respiring conditions is an aerated tank central part of waste water treatment plants as Fig. 2 reveals. From the aerated tank desaminated and decarburated waste water flows into increasingly installed nitrification-denitrification units for eliminating N-containing compounds more efficiently from the effluent. The CO_2 and ammonium enriched waste waters from the aerated tank is channeled through another aerated tank where autotrophic nitrifying bacteria find favorable conditions for conserving energy by oxidizing NH_4^+ to NO_3^- and forming biomass from CO_2. The waste water enriches with and flows in an unaerated tank supplemented with easily available electron donators for being returned favorably as N_2 into the atmosphere. The microbial biomass (sewage sludge) accumulating during waste water treatment is minimized in its volume in CH_4 and CO_2 producing fermenters (Fig. 2), and, if environmentally safe, composted and incorporated into agricultural soils.

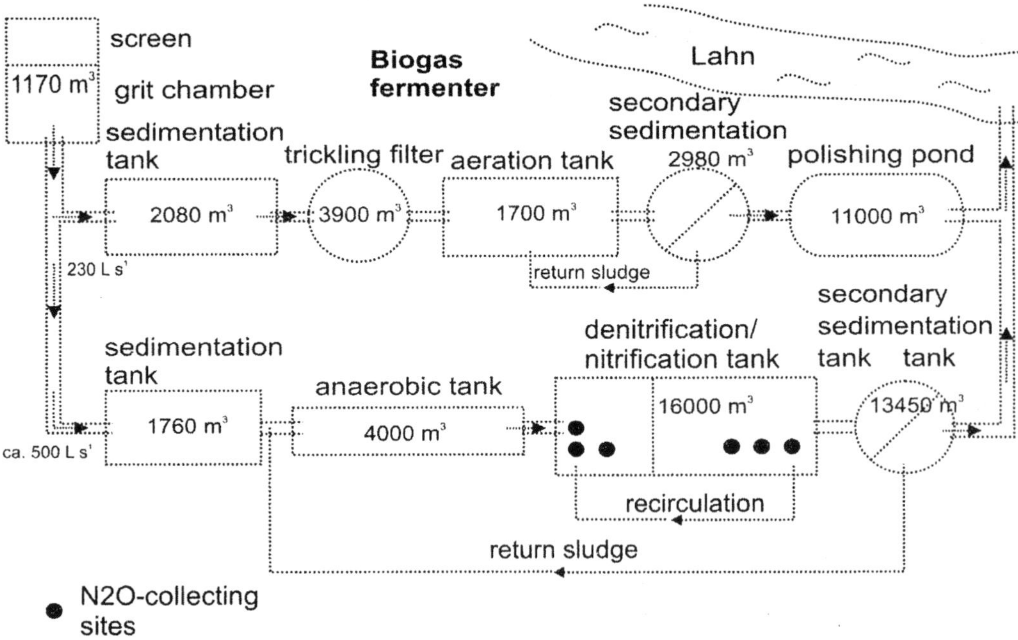

Fig. 2: Schematic design of the municipal waste water treatment plant of the City Giessen, Germany (details see text)

Microbial and Molecular Basis of Denitrification

The denitrifiers of Table 1 can taxonomically be subdivided into $\alpha-, \beta-, \gamma-, \varepsilon-$ proteobacteria, high and low GC Gram positive bacteria, Gram negative bacteria, pathogenic bacteria as *Brucella suis, B. melitensis* or *Mycobacterium bovis* having a nitric reductase, Archaea and fungi. Genome sizes of bacterial denitrifiers may range between 2.2 (*Neisseria gonorrhoeae*), 2.6 (*Pseudomonas lemoignei*), 3.7 to 4.6 (*P. stutzeri*), 5.9-6.5 (*P. aeruginosa*) 6.63 (*P. fluorescens*), 6.8 (*P. glathei*) and 8.7 Mb (*Bradyrhizobium japonicum*). Denitrifiers are competitors for the plant available The overall denitrification process is plant controlled by nutrient and O_2 competition, evapotranspiration, exudation and the offer of individually composed litter dead leaves, stem, root materials, (Chapman *et al.*, 2006). Additionally is the denitrification process in soils and waters controlled by the variables of Fig. 3. Essential regulators are nitrate, available carbon and the denitrifying microflora, because denitrification stops immediately if one of the 3 controlers is missing. Oxygen-free conditions favour denitrification, though gaseous N-species are continiously released even from well-aerated upland soils (Mosier and Parkin, 2006).

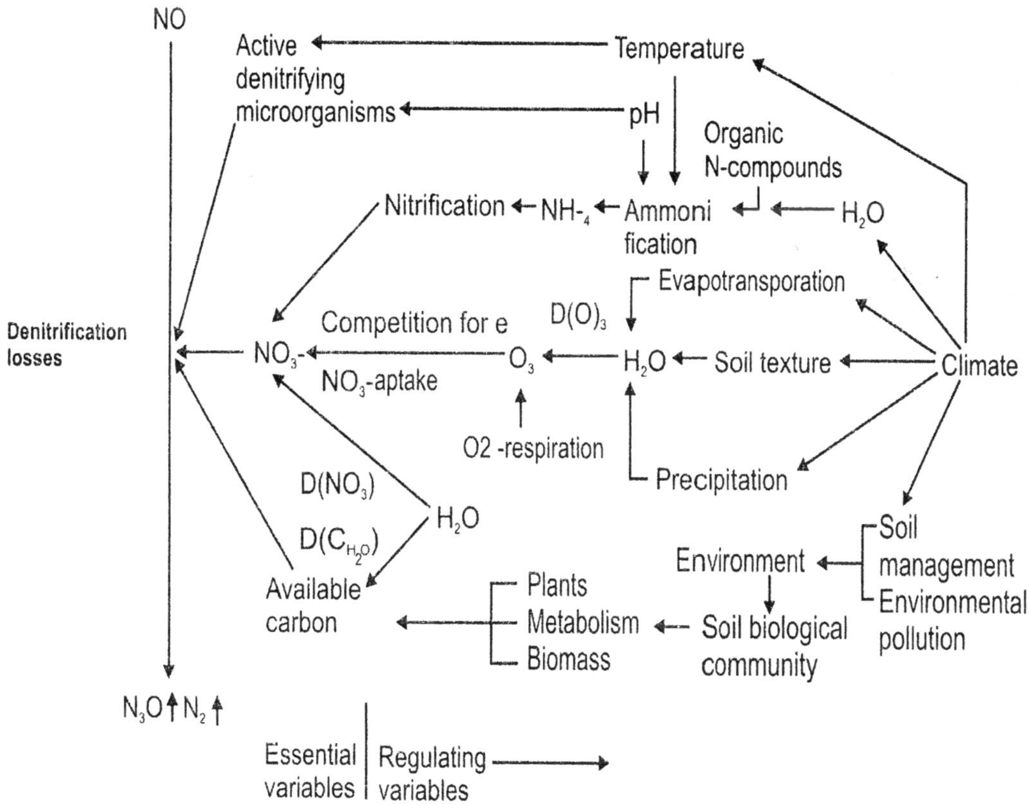

Fig. 3: Variables controlling denitrification in soils or constructed wetlands (D = diffusion; C_{H20} = water-soluble carbon)

The nitrate reductases of the facultative aerobe or anaerobe denitrifiers can be subdivided into a respiratory, a periplasmatic, an assimilatory and two dissimilatory nitrate reductase-types (Zumft, 1997). Denitrifiers have generally more than one of the three types of nitrate reductases available evidenced by the coexistence of all three nitrate reductases in *Ralstonia eutrophica* and *Paracoccus denitrificans.* The membrane-bound nitrate reductase (NarKC; Fig. 4) has been purified from a large variety of microorganisms and is only expressed under anaerobic growth conditions. In contrast to NarKC, the periplasmatic nitrate reductase (NapAB; Fig. 4) is actively synthesized only in the presence of oxygen. Both, Nar and Nap initiate an electron flow on nitrate through the quinone pool and cytochrome cyt b/c_1 ($NO_3^- + 2[H] \rightarrow NO_2^- + H_2O; \Delta G^{0'} -163\,kJ$). Nar is composed of a catalytic α-subunit (~140 kDa) encoded by *narG* containing a molybodopterin cofactor; a soluble -subunit (~60 kDa), encoded by *narH* containing four [Fe-S] clusters, and the -subunit (~20 kDA), encoded by *narl* containing two b-type haems. The catalytic sites of the dissimilatory nitrate reductase is oriented towards the cytoplasm and generates nitrite at the inner phase of the membrane. The periplasmatic nitrate reductase (Nap), only present in Gram-negative bacteria, is a heterodimer composed of the NapA catalytic subunit, containing a molybdopterin cofactor (90 kDa) and the NapB subunit containing two *c*-type haems. The NapAB complex is located in the periplasm. It receives an electron from the membrane-bound NapC containing 4 *c*-type haems. Nitrate-reducing and true denitrifying bacteria harbour either the membrane-bound, or the periplasmatic nitrate reductase, or both types of reductases. Isolates of a fluorescent Pseudomonad community showed *nar*G (56%), *nap*A (51%) and both 15.5% nitrate reductase genes (Philippot, 2005a). Genes encoding Nar are found in most of the bacterial divisions with 1-3 copies of the *nar*G gene per cell. The genome of *Streptomyces coelicolor* harbours obviously three *narGHJI* operons and that of *Burkholderia pseudomallei* two (Philippot, 2005a). Some of the bacteria, *e.g.*, *Escherichia coli,* reduce NO_3^- only to NO_2^- The assimilatory reduction of NO_3^- and NO_2^- to NH_3 fulfills biosynthetic purposes (Benckiser *et al.*, 2005).

Nitrite reductases in gram-negative bacteria are periplasmatic enzymes steered by the structural genes *nir*S and *nir*K (Fig. 4). Even non-denitrifiers as *Nitrobacter vulgaris* possess a nitrite reductase. The nitrite reductase of *Nitrobacter vulgaris* with Cu as a prosthetic metal and two subunits of 115 and 65 kDa releases NO and 4% ammonia. Nitrite is toxic to bacteria during the exponential growth phase in acidic environments and at concentrations greater 10-20 mM. The mechanism behind nitrite toxicity remains unclear. It is assumed that bacteria are susceptible to $NO_2^{\bullet}$, $NO_2^{\bullet}$, NO^+ and NO radicals with unpaired electrons as distinct chemical property. NO^+ and NO form spontaneously from nitrite favorably in acidic media and interact with metals and the bacterial enzymes to metal-nitrosyl complexes. NO with its nitrosating and deaminating reactivity is toxic to bacteria, fungi, microbial parasites, tumor cells, viruses and acts as mutagen to $C \rightarrow T$, $AT \rightarrow GC$, and $GC \rightarrow AT$ transitions. Under the frequently alternating aerobic and anaerobic living conditions of denitrifiers NO reoxidizes to nitrite ($4\,NO + O_2 + H_2O \rightarrow 4H^+ + 4\,NO_2^-$). Macrophages kill invading pathogens by attacking them with NO radicals. The pathogenity and survival of pathogenic denitrifies having the potential to reduce NO for energy gaining purposes may, in reverse, improve (Philippot, 2005b).

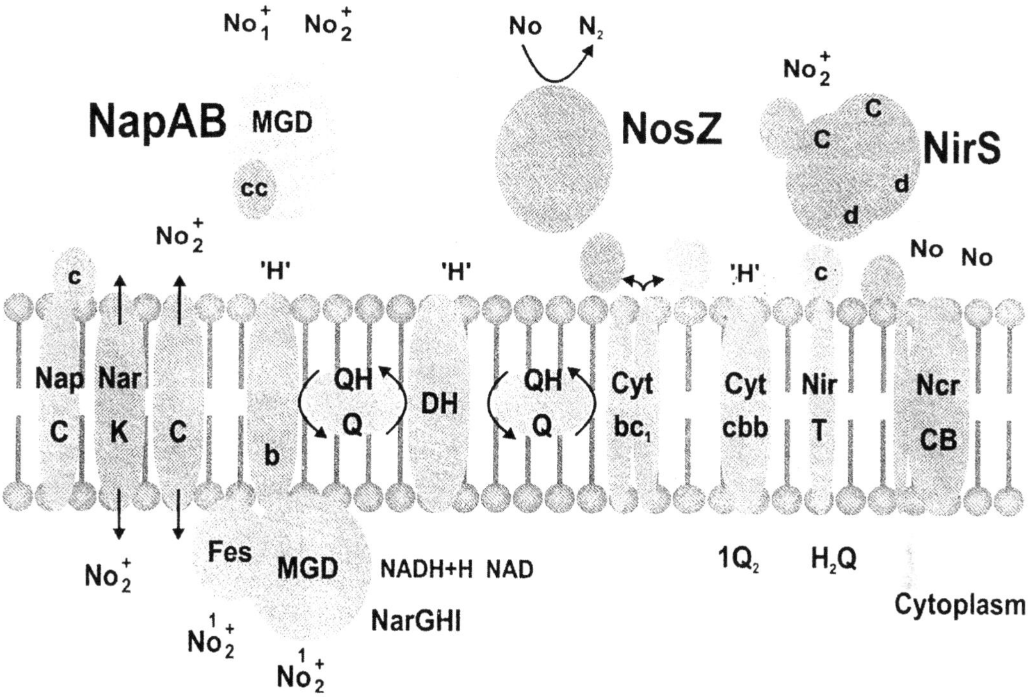

Fig. 4: Organization and sidedness of the anaerobic electron transfer chain of P. stutzeri (Zumft, 1997) DH = NADH dehydrogenase complex; Q, QH_2 = quinone cycle; Cyt bc_1 = cytochrome bc_1 complex; Cyt cb = cytochrome cb terminal oxidase complex; Nar = respiratory nitrate reductase; Nir = nitrite reductase; Nor = NO reductase; Nap = periplasmatic nitrate reductases; Nos = N_2O reductase soluble in the periplasmatic space; FeS = iron-sulfur centers; b, c, d_1 = hem B, C, D_1; cyt c = unspecified c-type cytochromes accepting lectrons from bc_1 complex and acting on N_2Or and Nor; cyt c_{551} = cytochrome c_{551}; AP = postulated nitrate/nitrite antiporter.

Assembling membrane-bound NO is quickly denitrified to N_2O (free energy, $\Delta G^{0'}$, 306, 3 kJ mol^{-1}). Dentrifying *Pseudomonas stuzeri* cultures stripped by a carrier gas release half to three quarters of the membrane-bound NO into the surrounding environment. Aerated (stripped) waste water tanks need in this respect attention (Stuven and Bock, 2001).

The chemolithoautotrophic obligate anaerobic anammox bacterium, Candidatus "*Kuenenia stuttgartiensis*" (anammox = anaerobic ammonium oxidation) denitrifies nitrate or nitrite with ammonium as electron donator to N_2 without releasing NO and N_2O ($3NO_3^- + 5NH_4^+ \rightarrow 4N_2 + 9H_2O + 2H^+; \Delta G^o = 297$ kJ mol^{-1} NH_4^+; Cirpus *et al.*, 2005). Both, the nitrate and NO reductase are linked to proton extrusion. The redox gradient of the NO_3^- / NO_2^- couple (+0.43 V) allows two proton translocations, one less than during aerobic respiration with O_2. The proton motive forces during the stepwise reduction of NO_3^- to N_2 create ATP by an ATPase in the usual fashion.

Fungal denitrification mostly ends at the N_2O level. *Taloromyces flavus, Fusarium solani* and *Cylindrocarpon tonkinense* make an exception. *Taloromyces flavus* releases preferently NO. Oxygen stressed *Fusarium solani* and *Cylindrocarpon tonkinense* reduce

NO_3^- until to N_2. The membrane-bound nitrate reductase of *F. oxysporum* and the nitrite reductase of *C. tonkinense* are mitochondria associated as sequenced 16 rDNA data exhibited. Mitochondria are bacteria derived eukaryotic cell endosymbionts and it can be speculated that fungi may have inherited the potential to denitrify from an engulfed ancient bacterium. N-containing compounds as azide and salichlhydroxamic acid react with fungal nitrite by yielding N_2 and N_2O.

The periplasmatic N_2O reductase (Fig. 4), purified from heterotrophic, phototrophic, and chemolithoautotrophic Gram positive and negative bacteria, converts respiratorily N_2O to N_2 ($N_2O + 2H^+ + 2e^-$ ® $N_2 + H_2O$; $DG^{0¢} = -$ 339.5 kJ mol^{-1}). In *P. denitrificans*, *Rhodobacter sphaeroides* and *Rhodobacter capsulatus* the cytochrome bc_1 complex transfers electrons on N_2O. The capability to reduce N_2O to N_2 is often lost as shown for *Clostridium nephridii, Pseudomonas chloroaphis,* a strain of *P. fluorescens, Roseobacter denitrificans, Thauera aromatica, Thauera selenatis, Azoarcus evansii,* or *Bacillus halodenitrificans. Wolinella succinogenes,* per definition a non-denitrifying bacteria reduces nitrate to nitrite, nitrite to ammonia, and N_2O to N_2. In *Ralstonia eutropha* and *Sinorhizobium mellioti* is the N_2O reductase plasmid-encoded and exists distinguished by redox and spectroscopic properties in several forms. The multicopper protein-N_2O reductase isolated from *P. stuzeri* and known as pink form sustains cell growth in the presence of copper and N_2O as sole electron acceptor. Other N_2O reductases use for the conversion of N_2O to N_2, *e.g.*, with benzene or alkylbenzenes as e^- donator Mo, Zn or Ni as transition metal ($M^n + N_2O \rightarrow O = M^{n+2} + N_2$). Other metalloenzymes, *e.g.* the nitrogenase, are evidently also able to catalyse the reduction of N_2O to N_2 (Zumft, 1997).

The NO_3^-—and nitrous oxide reductases have in the denitrification complex the largest degree of independence, while the membrane-bound nitrite and NO reductases are interdependently and hierachically regulated by a control circuit at both the transcriptional and the enzyme activity level, presumably to prevent the accumulation of NO radicals. FNR-like regulators (FNR = transacting protein of *E. coli* that activates gene expression under anaerobic conditions) coordinate the complete denitrification process (details, Zumft 1997). The O_2-sensing and signal transducing system used as main switch in genes expression during bacteroidal N_2-fixation of *Bradyrhizobium japonicum* regulates evidently also the denitrification process in *Bradyrhizobium japonicum.* This unveils that at least in azotrophs both processes are closer interwined as previously believed. The potential of N_2-fixing bacteria to denitrify provides substrate (N_2) and ATP by removing simultaneously nitrate which inhibits the nitrogenase complex (Benckiser *et al.*, 2005).

Gaseous nitrogen losses from soils and waste water treatment plants

Longterm studies in agricultural production systems document accelerated net losses of soil organic C and N (Mosier and Parkin, 2006). *In situ* quantifications of N_2O emissions and total denitrification losses from soils under natural vegetation, agricultural cropping systems and in waste water treatment plants reveal that soils as well as waste waters release continously not only N_2 but also N_2O into the atmosphere (Benckiser *et al.*, 1996; Fey *et al.*, 1999; Mosier and Parkin, 2006). As most important controllers of autotrophic and heterotrophic nitrification, nitrifierdenitrification and heterotrophic denitrification in no-till soils, conventionally cultivated soils and waste water treatment plants, soil moisture (oxygen), nitrate—and organic carbon availability have been identified. Exceeds the availability of the electron donator that of the electron acceptor nitrate/nitrite considerably then preferentiably NH_3 is released (Fig. 5). Carbon nitrate ratios in favor of the electron donator are found, *e.g.*, in the rumen of a bovine (Benckiser, 2006). More balanced carbon nitrate ratios or those in favour of the electron acceptor nitrate are rather typical for soils. Endproduct of soil NO_3^--respiration is mostly N_2 accompanied by NO and/or N_2O (Fig. 5). A shift towards N_2O is supported when O_2 concentrations in the soil atmosphere, *e.g.*, increase from 0 to about 5% (v/v). Under well-drained, aerobic soil conditions (WPFS 50% and below) NO, usually from nitrification, is the dominant gas. Under wetter conditions (WPFS 50-80%), where anaerobic microsites become increasingly abundant, N_2O from nitrification and denitrification is the dominant gas. Under very wet conditions with little oxygen diffusion but sufficient electron donators (WPFS > 80%), N_2 from denitrification is the dominant end product. These conclusions drawn from many *in-situ* observations hold true for most field conditions and are applicable across a diverse array of agricultural fields (Mosier and Parkin, 2006).

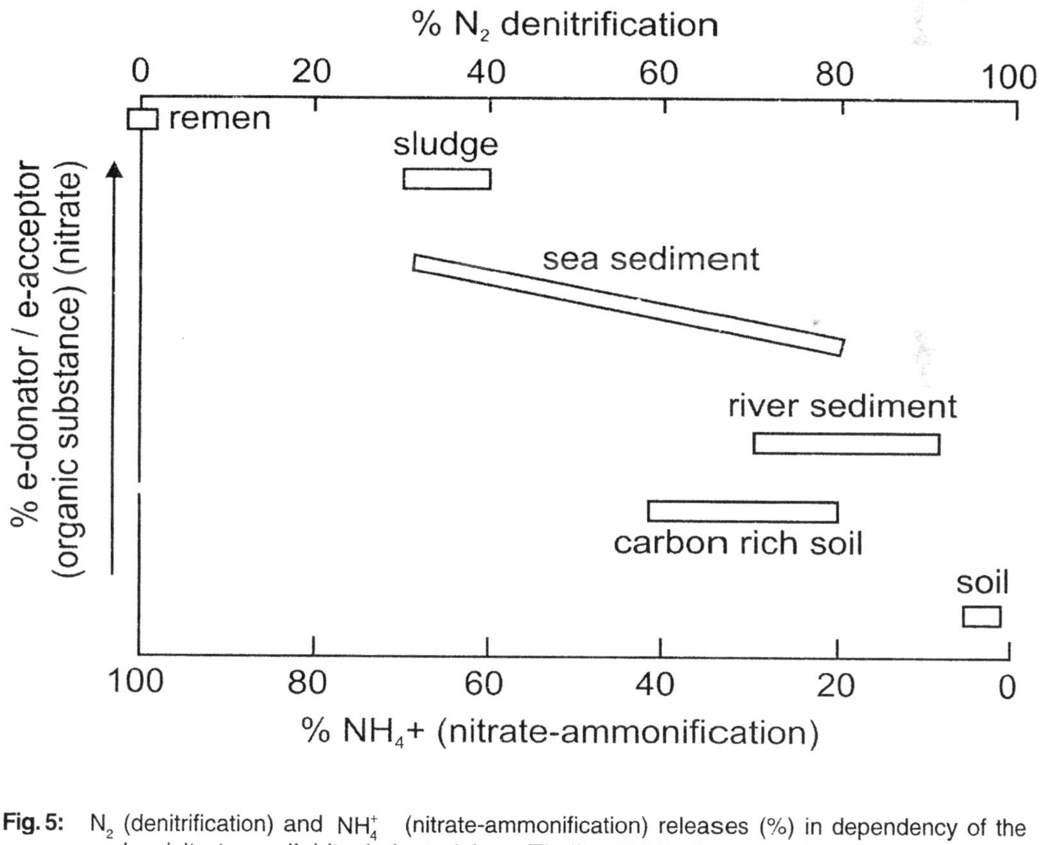

Fig. 5: N_2 (denitrification) and NH_4^+ (nitrate-ammonification) releases (%) in dependency of the carbon/nitrate availabity (adapted from Tiedje, 1988). Is more nitrate than reduction equivalents (NADH) available or when the oxygen partial pressure increases from 0 to about 5% (v/v) then the N_2O emissions are enhancing.

Agriculture with its livestock production and cropping systems, latter are fertilized with about 85 Tg N worldwide, adds annually about 5.6 Tg N_2O-N to the current atmospheric N_2O concentration of about 318 ppbv and is blamed to be a major contributor to emissions of the greenhouse gas N_2O. A rule of thumb is the greater the N-flow the greater the amount of N_2O emitted into the atmosphere. Consequently, a proper N_2O emission management in productive agricultural systems requires an understanding of the dominant factors affecting the ratios of the gaseous N-products. Managing concepts should include:

1. N availability has to be managed in a way that it seldom exceeds the plant demand.
2. N fertilizer placement has to consider gaseous diffusivity aspects.

Soil structure and water status determine gas diffusion in soils. Soil texture and porosity define drainage classes which take influence on the variables of Fig. 3 and determine the ratio between O_2-respiration and denitrification. Accordingly, agricultural soils should be managed in a way that oxygen can diffuse into deeper soil layers, and the denitrification gases NO and N_2O can reside long enough inside of the soil that they get the chance being further reduced to N_2 or NH_3. Fertilized perennial ryegrass pastures of Northern Ireland exhibit often WPFS > 70%, soil temperatures above 4°C and nitrate concentrations in the top 10 cm of the soil profile of above 2 mg kg^{-1}. Such conditions with denitrification rates as high as > 1 kg N ha^{-1} d^{-1} reveal that soil moisture, temperature and nitrate concentration provide useful rule of thumb for predicting broadly the N loss by denitrification in a particular field (Mosier and Parkin, 2006). Potentially important N_2O and NO emission events occur in rice-based agriculture mainly during the non-flooded times of the year and may total to 2.0-2.7% (N_2O) or 1.1-1.5% (NO) of the N applied by fertilization, respectively. Annual N_2O emissions from tropical upland cotton, maize, sorghum, and wheat production systems can range between about 8.5 and 56 kg N ha^{-1}. From the many N_2O emission measurements carried out under varying climatic conditions and N-availabilities it can be concluded that with an appropriate soil moisture management and the achievement of an organic matter nitrate ratio in favor of carbon availability the N_2O release from soils can be lowered (Fig. 5).

Our broadening knowledge about nitrification, nitrifer denitrification, assimilatoric nitrate ammonification and dissimilatory nitrate reduction (denitrification, nitrate ammonification) enters increasingly the waste water purification systems. Alone in Germany are annually about 0.4-0.5 Tg nitrogen channeled through waste water treatment plants. The modern municipal waste water treatment plant of the German City Giessen shown in Fig. 2 purifies about 15×10^6 m^3 waste water per year. From the inflowing 30-75 mg N L^{-1} waste water are about 92 to 98% eliminated and about 188 kg N released as N_2O (Benckiser *et al.*, 1996). Each tank of Fig. 2 contributed to the overall N_2O emission. Surprisingly, released the well and homogeneously aerated nitrification tank about 45 times more N_2O than the denitrification unit or 94 times more than the activated sludge unit. Of concern is not only the emitted but also the dissolved N_2O. Alone during the winter period 1995-1996 (November to March) discharged the waste water treatment plant of Fig. 2 an amount of 1005 kg N_2O-N dissolved in the effluent into the receiving river. Accordingly, dissolved N_2O-N should not be overlooked in the nitrogen balance.

Small communities and industries include increasingly decentralized constructed wetlands in their decisions how to purify economically waste water. Constructed wetlands, an example shows Fig. 6, avoid long distance transports of the waste water and are typically constructed by removing the soil body down to 60 cm and sealing the bottom and the sides with a water-tight foil. The excavated soil is mixed with peat and refilled. The soil-peat mixture forms a porous groundfilter generally planted with common reed (*Phragmites australis*). It is connected to a chamber ditch from which the waste water is received after sedimentation of the solid components. Planted soil-peat groundfilters regularly amended with waste water resemble highly organically manured agricultural soils. Wetlands as shown in Fig. 6 release as agricultural soils or the waste water treatment plant of Fig. 2 continously the greenhouse gas N_2O even during the winter months (November to March) at temperatures of –1.5°C in 10 cm depth of the groundfilter or 2°C at 50 cm depth. The average N_2O-N emission rate of the plant shown in Fig. 6 amounted to 15 mg N per inhabitant equivalent and day (Fey *et al.*, 1999).

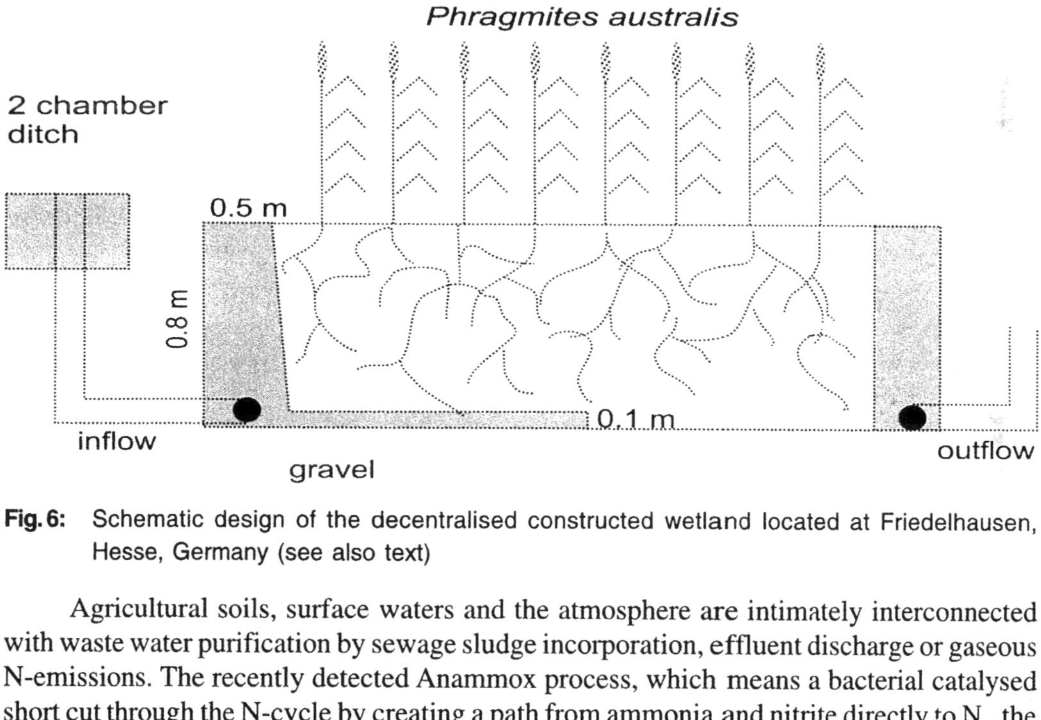

Fig. 6: Schematic design of the decentralised constructed wetland located at Friedelhausen, Hesse, Germany (see also text)

Agricultural soils, surface waters and the atmosphere are intimately interconnected with waste water purification by sewage sludge incorporation, effluent discharge or gaseous N-emissions. The recently detected Anammox process, which means a bacterial catalysed short cut through the N-cycle by creating a path from ammonia and nitrite directly to N_2, the newly discovered nitrate reduction pathway mediated by Fe(III)-reducing and Fe(II)-oxidizing microorganisms found in acetate-enriched wetland sediments, and methane oxidation in the presence of the redox pair NO_2^- / NO_3^- to CO_2 and N_2 by a methane oxidizing microbial consortium found in the anoxic sediment of a fresh water canal contaminated with agricultural run-off ($5CH_4 + 8NO_3^- + 8H^+ \rightarrow 5CO_2 + 4N_2 + 14H_2O$; $\Delta G^{0'} = -765$ kJ mol^{-1} CH_4; $3CH_4 + 8NO_2^- + 8H^+ \rightarrow 3CO_2 + 4N_2 + 10H_2O$; $\Delta G^{0'} = -928$ kJ mol^{-1} CH_4; Raghoebarsingh *et al.*, 2006) may add in future to improve the existing waste water treatment systems of Figs. 2 and 6.

References

Albrecht, A., Ottow, J.C.G., Benckiser, G., Sich, I. and Russow, R. 1997. Incomplete denitrification (NO and N_2O) from nitrate by *Streptomyces violaceoruber* and *S. nitrosporeus* revealed by acetylene inhibition and ^{15}N gas cromatography-quadrupole mass spectrometry analyses. Naturwissenschaften, ***84***: 145-147.

Benckiser, G. 2006. Principles behind order and sustainability in natural successions and agriculture. In: Biodiversity in agricultural production systems. Benckiser, G. and Schnell, S. (eds). CRC, Baton Rouge, USA.

Benckiser, G. 1997. Organic inputs and soil metabolism. In: Fauna in soil ecosystems. Benckiser, G. (ed.) p. 7-62. Marcel Dekker, New York.

Benckiser, G., Lorch, H.J. and Ottow, J.C.G. 1995. Quantification of total denitrification losses from undisturbed field soils by the acetylene inhibition technique. In: Methods in Applied Microbiology and Biochemistry. Alef, K. and Nannipieri, P. (eds). Academic Press, London, pp. 473-478.

Benckiser, G., Eilts, R., Linn, A., Lorch, H.J., Sümer, E., Weiske, A. and Wenzhöfer, F. 1996. N_2O emissions from different cropping systems and from aerated, nitrifying and denitrifying tanks of a municipal waste water treatment plant. *Biol. Fertil. Soils*, ***23***: 257-265.

Benckiser, G., Schneider, H., Hammeed, M.S., Kumar, K., Balanchandar, D. and Kannaiyan, S. 2005. Do rhizobial denitrification and nitrate ammonification act as counteracting tools for nitrate dependent inhibition of dinitrogen fixation. *Ind. J. Microbiol.*, ***45***: 195-203.

Benckiser, G. and Schnell, S. 2006. Biodiversity in Agricultural production Systems. CRC, Baton Rouge, USA.

Burth, I., Benckiser, G. and Ottow, J.C.G. 1982. N_2O-Freisetzung aus Nitrit (Denitrifikation) durch ubiquitäre Pilze unter aeroben Bedingungen. Naturwissenschaften, ***69***: 598-599.

Chapman, S.K., Langley, J.M., Hart, S.C. and Koch, G.W. 2006. Plants actively control nitrogen cycling; uncorking the microbial bottleneck. *New Phytologist*, ***169***: 27-34.

Cirpus, I.E., de Been, M., Op den Camp, H.J., Strous, M., Paslier, D.L., Kuenen, G.J. and Jetten, M.S. 2005. A new soluble 10kDa monoheme cytochrome c-552 from the anammox bacterium Candidatus "Kuenenia stuttgartiensis FEMS Microbiol Lett. ***252***: 273-278.

Fey, A., Benckiser, G. and Ottow, J.C.G. 1999. Emissions of nitrous oxide from a constructed wetland using a groundfilter and macrophytes in waste water purification of a dairy farm. *Biol. Fertil. Soils*, ***29***: 354-359.

Liz, J., Shaw, L.J., Graeme, W.N., Smith, Z., Fear, J., Prosser, J.I. and Baggs, E.M. 2006. *Nitrosospira* spp. can produce nitrous oxide via a nitrifier denitrification pathway. *Environ. Microbiol.* ***8***: 214-222.

Mosier, A.R. and Parkin, T.M. 2006. Gaseous Emissions (CO_2, CH_2, N_2O and NO) from Diverse Agricultural Production Systems. In: Biodiversity in Agricultural Production Systems. Benckiser, G. and Schnell, S. (eds). CRC, Baton Rouge, USA.

Philippot, L. 2005a. Tracking nitrate reducers and denitrifiers in the environment. *Biochem. Soc. Transact.* ***33***: 204-208.

Philippot, L. 2005b. Denitrification in pathogenic bacteria: for better or worst? *Trends Microbiol.* ***13***: 181-182.

Raghoebarsing, A.A., Pol, A., van de Pas-Schoonen, K.T., Smolders, A.J.P., Ettwig, K.F., Rijpstra, W.I.C., Schouten, S., Sinninghe Damste, J.S., Op den Camp, H.J.M., Jetten, M.S.M. and Strous, M. 2006. A microbial consortium couples anaerobic methane oxidation to denitrification. *Nature*. ***440***: 918-921.

Ratering, S., Benckiser, G. and Schnell, S. 2006. Metabolic diversity of microorganisms in agricultural soils. In: Biodiversity in Agricultural Production Systems. Benckiser, G. and Schnell, S. (eds). CRC, Baton Rouge, USA.

Shapleigh, J.P. 2005. The denitrifying prokaryotes. In: The Prokaryotes. Dworkin, M. (ed.) release 3.19 (3/18/2005).

Stuven, R. and Bock, E. 2001. Nitrification and denitrification as a source for NO and N_2O production in high-strength wastewater. *Water Res.* ***35***: 1905-14.

Thauer, R.K., Jungermann, K. and Decker, K. 1977. Energy conservation in chemotrophic anaerobic bacteria, *Bact. Rev.* ***41***: 100-180.

Tiedje, J.M. 1988. Ecology of denitrification and dissimilatory nitrate reduction to ammonium. In: Environmental Microbiology of Anaerobes. Zehnder, A.J.B. (ed.). John Wiley and Sons, New York, pp. 179-244.

Weber, K.A., Urrutia, M.M., Churchill, P.F., Kukkadapu, R.K. and Roden, E.E. 2006. Anaerobic redox cycling of iron by freshwater sediment microorganisms Environment. *Microbiol.* ***8***: 100-113.

Zumpft, W.G. 1997. Cell biology and molecular basis of denitrification. *Microbiol. Mol. Biol. Rev.* ***61***: 533-616.

Soil Microbiology and Soil Biotechnology : Vol. 1
Editor : S. Kannaiyan
Associated Publishing Co., New Delhi, 2010, pp. 76-88

6

Thiobacilli and the Environment: Extraction of Metals, Remediation of Pollutants and Amelioration of Soils

*K. Balaraman**

Introduction

Microorganisms have been known to form and decompose minerals in the earth's crust since geologically ancient times. The essential role of bacteria in the leaching of mineral ores was recognized in 1947 when *Thiobacillus ferroxidans* was identified as the organism principally responsible for the leaching of metal sulphide ores. Apart from *T. ferrooxidans* other iron-or sulfide/sulfur-oxidizing bacteria as well as archaea have been identified as potential agents for metal leaching. Sulphur-oxidizing microorganisms are also employed for desulphrification of coal and other fossil fuels. Microorganisms are also being used for the removal of heavy metals from the waste waters and sludges from industries and sewage, removal of abnoxious gases from industries, recycling of used nickel-cadmium batteries and amendment of polluted and sodic soils.

General characteristics of thiobacilli

There are several species of the genus *Thiobacillus* (Pseudomonadales). Thiobacilli are small, Gram-negative, rod-shaped cells (~0.5 × 1.0–4.0 μm) with some species motile by means of polar flagella (Anonymous, 1994). Energy is derived from the oxidation of one or more reduced sulphur compounds, including sulphides, sulphur, thiosulphate, polythionates and thiocyanate. Sulphate is the end product of sulphur compound oxidation, but sulphur, sulphite, or polythionates may be accumulated, sometimes, transiently, by most species. One species also derives energy from oxidizing ferrous iron to ferric iron. All species fix carbon dioxide by means of the Benson-Calvin cycle and are capable of autotrophic growth; some species are obligately chemolithotrophic, while others are also able to grow chemoorganotrophically. The genus includes obligate aerobes and facultative denitrifying types, and its species exhibit pH optima of 2-8 with temperature optima of 20-43°C. Distribution is seemingly ubiquitous in marine, freshwater, and soil environments, especially

* Vector Control Research Centre, ICMR, Pondicherry

where oxidizable sulphur is abundant (*e.g.*, sulphur springs, sulphide minerals, sulphur deposits, sewage treatment areas, and sources of sulphur springs, sulphide minerals, sulphur deposits, sewage treatment areas, and sources of sulphur gases, such as H_2S from sediments or anaerobic soils). Thiobacilli are generally resistant to a range of metal ions, with some variations between species and isolates within species. Normally, they are not found in large numbers but, if a local high concentration of a suitable substrate occurs, their activity becomes high resulting in a zone of pH 2.0 or lower.

Apart from several species of *Thiobacillus,* and other bacteria, archaea have also been observed to leach metals in nature. Important characteristics of some of these bacteria and archaea are in the Tables 1 and 2 (Anonymous, 1994).

Table 1: Characteristics of metal leaching bacteria

Microorganism	Nutrition	Growth Temp.	Growth pH
Thiobacillus ferrooxidans	Obligate chemolithoautotroph	30°–35°C	2.0–4.0
T. thiooxidans	Obligate chemolithoautotroph	25°–30°C	2.0–4.0
T. thioparus	Obligate chemolithoautotroph	25°–30°C	6.0–8.0
T. neopolitans	Obligate chemolithoautotroph	25°–30°C	6.0–8.0
T. denitrificans	Obligate chemolithoautotroph	25°–30°C	6.0–8.0
T. caldus	Facultative chemolithoautotroph	35°–45°C	0.5–5.5
T. novellas	Facultative chemolithoautotroph	23°–30°C	8.0–9.0
T. acidophilus	Facultative chemolithoautotroph	25°–30°C	1.0–6.0

Table 2: Characteristics of metal leaching archaea

Microorganism	Nutrition	Growth Temp.	Growth pH
Ferroplasma	Facultative chemolithotroph	33°–45°C	1.3–1.7
Sulfolobus	Obligate chemolithotroph	68°C	1.3–1.7
Metallosphaera	Facultative chemolithotroph	80°–85°C	1.0–4.5
Acidianus	Facultative chemolithotroph	70°–90°C	1.5–2.0

Although several species of *Thiobacillus* are seen in natural settings, for commercial purposes *T. ferrooxidans* has been used mostly. *T. ferrooxidans,* like other thiobacilli, derives the balance of its biosynthetic carbon requirements from CO_2 via Calvin-Benson cycle and carboxylation of phosphoenol pyruvate (DiSpirito, *et al.*, 1982). Oxidation of either ferrous iron or reduced sulphur compounds provides the energy required for CO_2 fixation and other functions in intermediary metabolism. The ferrous iron oxidation system of *T. ferrooxidans* is associated with the cellular membrane/envelope. The key iron-oxidizing enzyme appears to be Fe_2±cytochrome *c* oxidoreductase; cytochrome *a* and coenzyme Q are also associated with the process (Cobley & Haddock, 1975; Yates & Nason, 1966). A copper-containing protein, rusticyanin, has recently been identified. Rusticyanin reportedly serves as the initial electron acceptor upon oxidation of ferrous iron. The mechanism of sulphur-oxidation by *T. ferrooxidans* is similar to that of other thiobacilli. Elemental sulphur, tetrathionate, thiosulphate, and the sulphide moiety of varios non-iron minerals support growth of the organism; in all cases, sulphite (SO_3^{2-}) is the key intermediate molecule. Energy is produced from oxidation of sulphite to sulphate through the intareaction of the enzymes, sulphite oxidase, ADP sulphurylase, APS reductase, and adenylate kinase. An alternate sulphur-oxidation mechanism has been proposed in which elemental sulphur is oxidized to sulphite and then to sulphate via *T. ferrooxidans*, ferric iron reduction system (Sugio *et al.*, 1985). This alternative mechanism may be important under conditions of oxygen-limited growth.

Mechanism of leaching of metals

There are two school of thoughts concerning microbiologically assisted leaching of minerals namely, direct or indirect mechanism (Tributsch, 2001). In the indirect mechanism, the chemical attack by ferric iron or protons on a mineral sulfide results in the dissolution of the mineral and the formation of ferrous iron and various forms of sulfur. In the direct mechanism, attachment of microorganisms to the ore enhances the rate of mineral dissolution and direct attack is viewed as a process by which components within the bacterial membrane interact directly with the metal and sulfite moieties of the mineral by using an enzymatic type of mechanism (Devasia *et al.*, 1993). *T. ferrooxidans* (Hallmon *et al.*, 1992; Black *et al.*, 2001) has a strong affinity for mineral surfaces like pyrite to which it rapidly gets attached. The role played by the exopolysaccharide (EPS) produced by *T. ferrooxidans* (Rodriguez-leiva & Tributsch, 1988; Royas *et al.*, 1995; Schippers & Sand, 1999) has been emphasized when attachment to a mineral is accomplished. The EPS layer attaches the bacteria to a mineral surface and forms the matrix in which the cells divide and eventually form a biofilm. Iron species within the EPS layer have been estimated to be ~53 g/litre (Fowler *et al.*, 1999). The iron-impregnated EPS layer serves as a reaction space in which the high concentration of the ferric iron mounts chemical attack on the valence bonds of the mineral. In this process, the ferric iron is reduced to ferrous, which is reoxidised to ferric by the iron-oxidizing *T. ferrooxidans*.

$$14\,Fe^{2+} + 3.5\,O_2 + 14\,H^+ \rightarrow 14\,Fe_3^{3+} + 7\,H_2O.$$

This microbiologically mediated reoxidation of ferrous iron can result in a localized increase in pH within the EPS near the surface of the mineral and it is this local rise in pH that assists in mineral dissolution (Holmes *et al.*, 1999; Hutchins *et al.*, 1986). It has been reported that the oxidation of different metal sulfides proceeds via different intermediates therefore the dissolution reaction is not identical for all metal sulfides. A thiosulfate mechanism has been proposed for the oxidation of acid insoluble metal sulfides such as pyrite (FeS_2), molybderite (MoS_2) and tungterite (WS_2), and a polysulfide mechanism for acid soluble metal sulfides such as sphalerite (ZnS), chalcopyrite ($CuFeS_2$) and galena (PbS). In the thiosulfate mechanism, solubilization is through a ferric iron attack on the acid insoluble metal sulphides, with thiosulfate being the main intermediate and sulfate the main end product. Using pyrite as an example the reactions proposed (Fowler *et al.*, 1999) are:

$$FeS_2 + 6\,Fe^{3+} + 3\,H_2O \rightarrow S_2O_3^{2-} + 7\,Fe^{2+} + 6\,H^+$$
$$S_2O_3^{2-} + 8\,Fe^{3+} + 5\,H_2O \rightarrow 2\,SO_4^{2-} + 8\,Fe^{2+} + 10\,H^+.$$

In the case of polysulphide mechanism, solubilization of the acid-soluble metal sulfide is through a combination of ferric anions and proteins with elemental sulfur as the main intermediate. This elemental sulfur is relatively soluble but may be oxidized to sulfate by sulfur oxidizing bacteria. ZnS is an example of an acid soluble mineral.

$$MS + Fe^{3+} + H^+ \rightarrow M^{2+} + 0.5\,H_2Sn + Fe^{2+}\ (n \geq 2)$$
$$0.5\,H_2Sn + Fe^{3+} \rightarrow 0.125\,Sn + Fe^{2+}\ H^+$$
$$0.125\,S_8 + 1.5\,O_2 + H_2O \rightarrow SO_4^{2-} + 2\,H^+.$$

This explains why strictly sulfur oxidizing *T. thiooxidans* is able to leach metal sulfides but not others. The role of microorganisms in the solubilization of metal sulfides is therefore to provide sulfuric acid for a proton hydrolysis attack and to keep the iron in the oxidized ferric state or an oxidative attach on the mineral.

Leaching of iron, copper and uranium

Several studies confirmed the presence of *T. ferrooxidans* in copper leaching operations (Brierley, 1977). After the top of the lower grade ore dump is levelled, leach solution is flooded or sprayed and this is followed by the colonization of *T. ferrooxidans* mainly in the top meter or so. The temperature may reach 90^{0}C in the center of the dump owing to the exothermic nature of leaching reactions. Leach solutions containing 0.5 to 2.0 g Cu liter^{-1} exit at the base of the dump and are conveyed to a central recovery facility. The following reaction then occurs (Brierley, 1978):

$$CuSO_4 + Fe^0 \rightarrow FeSO_4 + Cu^0$$

The maximum rate of copper dissolution by *T. ferrooxidans* was about 215 mg/liter/hr and the final copper concentration was as high as 55 g/liter (Sakaguchi & Silver, 1976).

Uranium leaching, using *T. ferrooxidans* involves periodic spraying or flooding of worked-out stopes and tunnels of underground mines with water or dilute sulfuric acid solution. Bacterial activity is limited to oxidation of pyrite and ferrous iron because *T. ferrooxidans* does not directly interact with Uranium minerals. Uranium leaching proceeds as follows (Brierley, 1980; Brierley, 1974):

$$UO_2 + Fe_2(SO_4)_3 + 2\,H_2SO_4 \rightarrow UO_2(SO_4)_3^{4-} + 2\,FeSO_4 + 4\,H^+.$$
$$UO_3 + 3\,H_2SO_4 \rightarrow UO_2(SO_4)_3^{4-} + H_2O + 4\,H^+.$$

Through microbiological oxidation of ferrous iron, uranium extraction as high as 68% has been realized during nine days of treatment with *T. ferrooxidans* (Guay *et al.*, 1977).

Leaching of cobalt, nickel, zinc, molybdenum, lead and cadmium

The leaching of cobalt, nickel and zinc by *T. ferrooxidans* is by indirect mechanism

(Brierley, 1977): Sulfuric acid is generated during sulfur oxidation (not metal sulphide oxidation) which solubilizes the metal. Leaching of lead by *T. ferrooxidans* is in effect an enrichment of lead in the metal matrix, *i.e.*, during the oxidation of PbS. The lead is converted to insoluble lead sulfate and other soluble metal sulfates are leached, leaving a concentrated lead product (Torma & Bosecker, 1982). *T. ferrooxidans* had been observed to leach metals from cobalt sulphide (Torma, 1973, 1978; Groudev, 1981; Brierley, 1983), gallium sulphide (Iyaletolinov *et al.,* 1977), nickel sulphide (Torma, 1973; Groudev, 1981; Brierley, 1983), lead sulphide (Torma & Guay, 1976; Fridman & Savari, 1983), antimony sulphide (Lawrence & Bruynesteyn, 1983) and zinc sulphide (Torma, 1973, 1978); Torma & Guay, 1976; Robinson, 1983).

Recovery of gold and silver

Gold ores are encapsulated by iron sulfide minerals and therefore are unavailable for contact with the cyanide. Pretreatment of gold ores with microorganisms degrade the matrix of iron sulfide minerals, allowing contact between the cyanide and the gold. Ores from which gold recovery was less than 50% with cyanide treatment alone yielded greater than 90% recovery when pretreated with *T. ferrooxidans* (Ashmead, 1955; Yudina *et al.,* 1979; Pooley, 1982; Ehrlich, 1985). An analogous process exists for silver ores that are refractory owing to encapsulation of the precious metal by iron sulfide minerals (Dugan & Apel, 1978; Yudina *et al.,* 1979; Pooley, 1982; Ehrlich, 1985). A mixed culture of *T. ferrooxidans* and *T. thiooxidans* can accumulate silver, amounting to upto 25% of the cell dry weight from the mineral acanthite (Ag_2S) (Detz & Barvinchak, 1979).

Coal desulfurization

Combustion of coal results in the formation of sulfur dioxide, an atmospheric pollutant. Sulfur in coal consists of minerals sulfides (primarily pyrite and marcasite), organic sulfur compounds, and sulfates. Inorganic sulfur generally predominates in high sulfur coals. Hence the reduction of inorganic sulfur content is often essential to render coal environmentally acceptable for combustion when used in conjunction with minimal emission-control processes. The use of acidophilic iron—and sulfur-oxidizing microorganisms for coal desulfurization was first proposed by Ashmead (1955) and Dugan & Apel (1978) inoculated an aqueous slurry of coal containing 3.1% pyrite sulfur with a mixed culture of *T. ferrooxidans* and *T. thiooxidans* and could remove ca. 97% of the pyrite sulfur in five days. Pyrite dissolution under thermophilic conditions is generally faster than that at near ambient temperature with *T. ferrooxidans* (Gokcay & Yurteri, 1983). A preliminary economic analysis has shown that microbial desulfurization compares favourably with alternative chemical coal cleaning methods and pyrite is more easily separated from coal during froth floatation following an initial exposure to *T. ferrooxidans* (Hedrick *et al.,* 1982; Pooley & Atkins, 1983; McCready & Legallais, 1984). Presumably pyrite is rendered more hydrophilic by selective attachment of the bacteria.

Remediation of environmental pollutants

Removal of sulphide waste waters

Sodium hydroxide solutions are used in petroleum refining to remove hydrogen sulphide and mercaptans from various hydrocarbon streams. The resulting sulphide laden waste stream is called spent sulphide caustic. Several studies indicate that these sulphide-rich waste waters could be treated with microorganisms.

An aerobic enrichment culture developed using a gas mixture of H_2S and methyl mercaptan (MeSH) as the sole energy source has been used to treat the spent sulphidic caustic containing both inorganic sulphide and mercaptans (Visser *et al.*, 1997) and complete oxidation of both inorganic and organic sulphur to sulphate was achieved. Chemostat cultures of a new *Thiobacillus* sp. had a very high sulphur forming capacity, converting almost all hydrogen sulphide in the medium to elemental sulphur. In another study, using *T. denitrificans* more than 800 m^3/d of waste water containing 100 mg/L sulfide and total dissolved solids of 4800 mg/L were successfully treated in an earthern pit (3000m^3) over a six month period (Sublette *et al.*, 1998) which resulted in complete removal of sulphides and elimination of associated odours (Table 3).

Table 3: Removal of environmental pollutants by thiobacilli

Source and Pollutant(s)	Thiobacilli used	Percentage removal	Reference
Sulphide waste (Petroleum refinery)	*T. denitrificans*	Sulphides & odours: 100	Sublette *et al.*, 1998.
Sulphuric pickling liquors (Steel industry)	*T. ferrooxidans*	Waste iron: 99.8	Garcia *et al.*, 1994.
Water waste (Animal husbandry)	*T. thioparus* (H_2S) & *Nitrosomonas europaea* (NH_3)	H_2S & NH_3: > 95	Chung *et al.*, 2000.
Water waste (Animal husbandry)	*Thiobacillus*	H_2S & CS_2: > 100	Hartikainen *et al.*,
Municipal sewage Sludge	*T. thioparus* & *T. thiooxidans*	Cadmium:83-96; Chromium:16-54; Copper:85-87; Manganese:91-94; Nickel:78-79; Lead:28-46; Zinc:82-96	Blais *et al.*, 1992.
Sludge (Municipal sewage)	*T. ferrooxidans*	Manganese:80; Zinc:80; Copper:20; Aluminium:10; Titanium:0.2	Lombardi & Garcia Jr., 2002
Industrial wastes	*T. ferrooxidans*	Iron, Manganese & Aluminium: 76-78	Solosio *et al.*, 2002.
Fly ash & lime (Municipal incinerator)	*T. ferrooxidans*	Lead (1594-5038 mg/kg): 40-69. Zinc (8273-16873 mg/kg): 168-80. Copper (495-465 mg/kg): 30-68. Cadmium (129-267 mg/kg): 92-95.	Mericer *et al.*, 1999.
Spent nickel-cadmium batteries	*T. ferrooxidans*	Cadmium: 100; Nickel: 97; Iron: 95	Cerruti *et al.*, 1998.

Through oxidation of the ferrous sulphate contained in steel industry sulphuric pickling liquors with *T. ferrooxidans,* a process involving recovery of iron oxides (alpha Fe_2O_3 type) was developed (Garcia *et al.*, 1994). The process yielded 50 Kg of Fe_2O_3 per m^3 of waste pickling liquor at 700^{0}C with 99.8% waste iron recovery (Table 3). In another study, an aerobic enrichment culture was immobilized in DuPont BIO-SEP beads, and used to treat a refinery spent-sulfidic caustic in a continuous flow, fluidized-bed column bioreactor (Conner *et al.*, 2000) which resulted in complete oxidation of both inorganic and organic sulfur to sulfate.

A gas mixture of H_2S and NH_3 generated from the animal husbandry and anaerobic waste water lagoons were treated using a mixture of specific mocroflora *viz., T. thioparus* for the removal of H_2S and *Nitrosomonas europaea* for the removal of NH_3 (Chung *et al.*, 2000). When a bio-filter packed with co-immobilized cells was continuously supplied with H_2S and NH_3 gas the efficacy of gas removal was >95% (Table 3). And with a peat bio-filter inoculated with *Thiobacillus* which in acidic environment, it was possible to remove H_2S and CS_2 simultaneously (Hartikainen *et al.*, 2001).

T. denitrificans was used to oxidize H_2S with concomitant reduction of N_2O to N_2 (Dalsgaard & Bak, 1992). In batch culture, oxidation of H_2S was stepwise, with transient accumulation of elemental sulfur and the final oxidation product was SO_4. In gradient culture, there was no notable accumulation of elemental sulfur and microsensor measurements of H_2S and N_2O showed that H_2S was oxidized directly to SO_4.

In order to remove high concentrations of H_2S from anaerobic waste water treatments in livestock farming, a novel process was evaluated for H_2S gas abatement involving the combination of chemical absorption and biological oxidation processes (Chung *et al.*, 2003). The results showed that the chemical process removed up to 300 ppm of H_2S, and nearly no acclimation time was required. The optimal operating conditions for the biological oxidation process were below pH 2.3 and 35ºC in which more than 90% Fe^{3+} + formation ratio was achieved. Interestingly, the optimal glucose concentration in the medium was 0.1% which favored Fe^{2+} oxidation and the growth of *T. ferrooxidans*.

Removal of heavy metals from municipal sewage sludge

The use of sewage sludge in agricultural land as a means of sludge disposal and recycling has been shown to be economical and suitable because of the presence of nutrients such as nitrogen and phosphorous (Lombardi & Garcia Jr., 1999). However, heavy metal concentration in sludge is one of the obstacles for the application of sludge on land (Aralp *et al.*, 2001). Apart from the presence of heavy metals, municipal sludges often contain other compounds that must be removed for its safe use in agricultural soils. Although there are various methods available, microbiological removal of heavy metals from municipal sludge is an outstanding option because of high metal solubilization rates and the low cost.

A mixed culture of two fast growing strains (*T. thioparus* and *T. thiooxidans*) was used to solubilize heavy metals from municipal sewage sludge (Blais *et al.*, 1992). The former strain decreases the initial sludge pH of 7.0–8.5 to 4.0–4.5 and then the latter strain begins growing and further reduces the pH to < 2.0. The rapid reduction of sludge pH through sulfur oxidation into sulphuric acid solubilizes the toxic metals as indicated below: cadmium: 83–96%, chromium: 16–54%, copper: 85–87%, manganese: 91–94%, nickel: 78–79%, lead: 28–46% and zinc: 82–96% and these levels are recommended for intensive use of residual sludge in agriculture (Table 3). Metal solubilization from the sewage sludge was better accomplished if additional energy source in the form of S_2 or Fe (II) is supplementd to the microbes, *T. thiooxidans* or *T. ferrooxidans* (Lombardi *et al.*, 2001) and among these *T. thooxidans* performed better than *T. ferrooxidans*.

The chemical fractionation and leaching of manganese, aluminium, zinc, copper, and titanium in municipal sewage sludge were investigated using *T. ferrooxidans* as leaching microorganism (Lombardi & Garcia Jr., 2002). As a result of bacterial activity, oxygen reduction potential increase and pH reduction were observed; the solubilization efficiency approached ca. 80% for manganese and zinc, 24% for copper, 10% for aluminium and 0.2% for titanium (Table 3). Metal solubilization was accomplished only when supplemented with Fe(II) as energy source. In another study (Aralp *et al.*, 2001) sulfur-oxidizing bacteria was applied to sludges generated from the co-treatment of municipal waste water and leachate, for the removal of selected heavy metals. The highest chromium, nickel and iron solubilization was obtained from the set containing 3% leachate and sulphur-oxidizing bacteria were totally inhibited in set containing 10% of leachate.

Removal of zinc and aluminium from industrial waste sludges

Biological solubilisation of heavy metals contained in two different kinds of industrial wastes was performed employing a strain of *T. ferrooxidans* (Solosio *et al.*, 2002). The wastes tested were a dust coming from the iron-manganese alloy production in an electric furnace (sludge-1) and a sludge coming from a process treatment plant of aluminium anodic oxidation (sludge-2). The experimental results pointed out the activity of *T. ferrooxidans* to maintain an environment (that initially has a pH of about 8) at strongly acid conditions (pH 2.5 to 3.5), producing sulphuric acid which is responsible for the solubilisation of metals. At wastes' initial concentration of 1%, the percentage of solubilised metals was 76 and 78, respectively for the sludge-1 and sludge-2 (Table 3).

In another study, the significance of pH in bioleaching of heavy metals from contaminated sediment was evaluated at different solid contents of sediments in a bench-scale reactor (Chen & Lin, 2001). The results showed that a temporal change of pH in the bioleaching process was effected by the buffering capacity of the sediment particulates. It was observed that solubilization of heavy metals from sediments is highly pH-dependent.

Removal of heavy metal(s) from the Fly ash from municipal incinerator

The purification of incinerator gases produces large quantities of fly ash and used lime (a contamination referred to as "air pollution control residues" (APCR). Fly ash as well as used lime contain levated levels of metals (Mercier *et al.*, 1999; Cerruti *et al.*, 1998). Biological solubilization of fly ash and used lime was carried out utilizing *T. ferooxidans* to render them non-hazarduous. The removal of lead from the APCR was 40-69 for APCR containing 1594-5038 mg lead/kg. Zinc removal varied from 68 to 80% for APCR containing 8273-16873 mg zinc/kg whereas copper was removed to the extent of 30-68% for APCR containing 495-465 mg copper/kg APCR. Cadmium was removed to the greatest degree, *i.e.*, 92-95% for APCR containing 129-267 mg cadmium/kg. After biological treatment, the leachate from the "TOP" (The toxicity characterization leaching procedure of the USEPA) was within the acceptable criteria of the USEPA (Table 3).

Removal of heavy metal(s) from spent nickel cadmium batteries

An economic and environmentally friendly process is necessary to recycle dry batteries.

Microbial leaching is one of the few techniques applicable for the recovery of the toxic metals from spent batteries and its principle is the microbial production of H_2SO_4 and simultaneous leaching of metals.

Bioreactors with *T. ferrooxidans* attached on elemental sulphur reached a maximum H_2+ production of 80 m mol kg^{-1} d^{-1} of support (Cerruti *et al.,* 1998). This medium when used for the indirect dissolution of spent nickel cadmium batteries, recovered 100% of cadmium, 96.5% of nickel and 95% of iron after 93 days. Recoveries higher than 90% were reached when anodic and cathodic materials were directly added to *T. ferrooxidans* culture with sulfur as the sole energy source.

In another study a system was developed which consisted of a bioreactor, settling tank and leaching reactor to leach metals from nickel-cadmium batteries (Zhu *et al.,* 2003). Indigenous thiobacilli, proliferated by using nutritive elements in sewage and elemental sulfur as substrates, were employed in the bioreactor to produce H_2SO4. The results showed that this system is efficient in leaching metals from nickel-cadmium batteries and that the sludge drained from the bottom of the settling tank could satisfy the requirements of EPAS regarding agricultural use.

Reclamation of sodic saline soil

Excess of salts in soils reduces plant development, increases the energy required for nutrients and water absorption, and requires a biochemical adjustment for the survival in stress conditions (Rhoades *et al.,* 1992). The efficiency of corrective substances such as gypsum and organic matter frequently used in the amendment of sodic soils has been evaluated in several studies. The chemical correctively used in saline and alkaline soils displace sodium and other ions of the soil exchange complex, and these cations leach from the soil profile (Miyamoto & Enriquez, 1990; Armstrong & Tanton, 1992). When elemental sulphur is used, a biological reaction takes place in soil, carried out by different species of *Thiobacillus,* producing sulphuric acid that reduces soil pH (Garcia Jr., 1992). In a soil solution, the dissociated hydrogen ion can displace the sodium from clay particles thereby promoting its subsequent leaching, reducing the soil salinity (Rupela & Tauro, 1973).

A green house experiment carried out with the objective of evaluating the effects of the elemental sulfur inoculated along with *Thiobacillus* as compared with gypsum in the amendment of a alluvial sodic saline soil (Stamford *et al.,* 2002). The treatments consisted of two levels of sulfur (300 and 600 kg/ha) and gypsum (1200 and 2400 kg/ha). Fifteen days after application of sulfur (1200 kg/ha) along with *Thiobacillus* promoted leaching of salts, reduced the concentration of exchangeable sodium of the soil (from 303 mg/l to 112 mg/l) and reduced the Electrical Conductivity to levels below that adsorbed in soil classified as sodic or saline sodic (4 dS/m). This treatment also reduced the level of exchangeable potassium and calcium. Compared to gypsum treatment the combination of sulphur and *Thiobacillus* was more efficient.

Conclusion

In conclusion it can confidently be said that the thiobacilli have got the potential to be exploited for alleviating enormous burden of environmental pollution although they have been in commercial use for extraction of minerals and metals from ores.

References

Anonymous. 1994. Aerobic chemolithotrophic bacteria and associated organisms, pp. 427-455. In: Group 12 of the Bergey's Manual of determinative Bacteriology (9^{m} Ed), Hensyl, W.R. ED., Pub. By Williams & Wilkins, Baltimore, Maryland, USA, 1994, pp. 1-787.

Aralp, L.C., Erdincler, A. and Onay, T.T. 2001. Heavy metal removal from waste water and backwater treatment sludge by sulphur oxidizing bacteria. *Water Sci. Technol.* **44**(10): 53-58.

Armstrong, A.S.B. and Tanton, T.W. 1992. Gypsum applications to aggregated saline-sodic clay topsoils. *J. Soil Sci.*, **43**: 249-260.

Ashmead, D. 1955. The influence of bacteria in the formation of acid mine waters. *Colliery Guardian,* **190**: 690-98.

Blake, R.C., Sasaki, K. and Ohmura, N. 2001. Does aporusticyanin mediate the adhesion of *Thiobacillus ferrooxidans* to pyrite? *Hydrometallurgy,* **59**: 357-72.

Blais, J.F., Auclair, J.C. and Tyagi, R.D. 1992. Cooperation between two *Thiobacillus* strains for heavy-metal removal from municipal sludge. *Con. J. Microbiol.* **38**(3): 181-187.

Brierley, C.L. 1974. Molybodenum leaching: use of high temperature microbe. *J. Less-Common Met.* **36**: 237-247.

Brierley, C.L. 1977. Thermophilic microorganisms in extraction of metals from ores. *Dev. Ind. Microbiol.* **18**: 273-284.

Brierley, J.A. 1978. Thermophilic ironoxidizing bacteria found in copper leaching dumps. *Appl. Environ. Microbiol.* **36**: 523-25.

Brierley, C.L. 1980. Leaching of chalcopyrite ore using *Sulfolobus* species. *Dev. Ind. Microbiol.,* **24**: 435-44.

Brierley, C.L. 1983. The prospects for using biotechnology for cobalt and nickel recovery from copper ores and solid waste materials. *Proc. Symp. Hydrometall. Copper Byprod. Rarer Met., Dallas,* pp. 7-11, New York, AIME, 113 pp.

Cerruti, C., Curutchet, G. and Donati, E. 1998. Bio-dissolution of spent nickel cadmium batteries using *Thiobacillus ferrooxidans. J. Biotechnology,* **62**(3): 209-219.

Chung, Y.C., Huang, C., Tseng, C.P. and Pan, J.R. 2000. Biotreatment of H_2S—and NH_3—containing waste gases by co-immobilized cells biofilter. *Chemosphere*, **41**(3): 329-336.

Chung, Y.C., Ho, K.L. and Tseng, C.P. 2003. Hydrogen sulfide gas treatment by a chemical biological process: chemical absorption and biological oxidation steps. *J. Environ. Sci. Health B.,* **38**(5): 663-79.

Cobley, J.C. and Haddock, B.A. 1975. The respiratory chain of *Thiobacillus ferrooxidans*: the reduction of cytochromes by Fe^{2+} and the preliminary characterization of rusticyanin a novel "blue" copper protein. *FEBS Lett.,* **60**: 29-33.

Conner, J.A., Beitle, R.R., Duncan, K., Kolhatkar, R. and Sublette, K.L. 2000. Biotreatment of refinery spent-sulfidic caustic using an enrichment culture immobilized in a novel support matrix *Appl. Biochem. Biotechnol.,* **84-86**: 707-19.

Dalsgaard, T. and Bak, F. 1992. Effect of acetylene on nitrous oxide reduction and sulfide oxidation in batch and gradient cultures of Thiobacillus denitrificans. *Appl. Environ. Microbiol.,* **58**(5): 1601-8.

Detz, C.M. and Barvinchak, G. 1979. Microbial desulphurization of coal. *Min. Congr. J.*, **65**: 75-86.

Devasia, P., Natarajan, K.A., Sathyanarayana, D.N. and Ramananda Rao, G. 1993. Surface chemistry of

Thiobacillus ferrooxidans relevant to adhesion on mineral surfaces. *Appl. Environ. Microbiol.*, **59**: 4051-56.

Dispirito, A.A., Silver, M., Voss, L. and Tuovinen, O.H. 1982. Flagella and pili of iron-oxidizing thiobacilli isolated from a uranium mine in Northern Ontario, Canada. *Appl. Environ. Microbiol.*, **43**: 1196-1200.

Dugan, P.R. and Apel, W.A. 1978. Microbiological desulfurization of coal. See Ref. 113, pp. 223-50.

Ehrlich, H.L. 1985. *Bioleaching of silver from a sulfide ore by continuous feeding of lixiviant.* Presented at Ann. Meet. Am. Soc. Microbiol., 85th, Las Vegas.

Fowler, T.A. and Crundwell, F.K. 1999. Leaching of zinc sulfide by Thiobacillus ferrooxidans: bacterial oxidation of the sulfur product layer increases the rate of zinc sulfide dissolution at high concentrations of ferrous ions. *Appl. Environ. Microbiol.*, **65**(12): 5285-92.

Fridman, I.E. and Savari, E.E. 1983. Treating carbon containing Ag-Au-As concentrates. *World Min.*, **36**: 45-47.

Garcia Jr., O. 1992. O enxofre e suas transformacoes microbians. In: Cardoso, E.J.B., S.M. Tsai, and M.C.P. Neves (Eds), Microbiologia do solo. Sociedade Brasileira de Ciencia do solo, Campinas, pp. 1-15.

Garcia, F.J., Rubio, A., Sainz, E., Gonzalez, P. and Lopez, F.A. 1994. Preliminary study of treatment of sulphuric pickling water waste from steelmaking by bio-oxidation with T*hiobacillus ferrooxidans. FEMS Microbiol. Rev.* **14**: 397-404.

Gokcay, C.F., Yurteri, R.N. 1983. Microbial desulphurization of lignites by a thermophilic bacterium. *Fuel,* **62**: 1223-24.

Groudev, S.N. 1981. Leaching of cobalt from synthetic cobalt sulphide by *Thiobacillus ferrooxidans* and *Thiobacillus thiooxidans. C.R. Acd. Bulg. Sci.*, **34**: 217-220.

Guay, R., Silver, M. and Torma, A.E. 1977. Ferrous iron oxidation and uranium extraction by Thiobacillus ferrooxidans. *Biotechnol. Bioeng.*, **19**(3): 727-40.

Hallmonn, R., Friedrich, A., Koops, H.P., Pommerening-Rosin, A., Rohda, *et al.*, 1992. Physiological characteristics of *Thiobacillus ferrooxidans* and *Leptospirillum ferrooxidans* and physio-chemical factors influence microbial metal leaching. *Geomicrobiol. J.*, **10**: 193-206.

Hedrick, H.G., Fernandes, J.B., Arrowood, M.J. and Franklin, T.J. 1982. Desulfurization of coal by biological pretreatment, state-of-the-art. *Dept. Nat. Resour. Interim. Rep.*, Louisiana Tech. Univ., Ruston. p. 56.

Holmes, P.R., Fowler, T.A. and Crundwell, F.C. 1999. The mechanism of bacterial action in the leaching of pyrite by *Thiobacillus ferroxidans. J. Electrochrom. Soc.*, **146**: 2906-12.

Hutchins, S.R., Davidson, M.S., Brierley, J.A. and Broerley, C.C. 1986. Microorganisms in reclamation of metals. *Ann. Rev. Microbiol.*, **40**: 311-336.

Ilialetdinov, A.N., Kamalov, M.R. and Stukanov, V.A. 1977. Microbiological leaching of zinc and lead ores of the Tekeli deposit. *Mikrobiologiia,* **46**(5): 857-66.

Lawrence, R.W. and Bruynesteyn, A. 1983. Biological pre-oxidation to enhance gold and silver recovery from refractory pyretic ores and concentrates. *CIM Bull.*, **74**: 107-10.

Lombardi, A.T. and Garcia Jr., O. 1999. An evaluation into the potential of biological processing for the removal of metals from sewage sludges. *Crit. Rev. Microbiol.* **25**(4): 275-288.

Lombardi, A.T. and Garcia Jr., O. 2002. Biological leaching of Mn, Al, Zn, Cu and Ti in an anaerobic sewage sludge effectuated by *Thiobacillus ferrooxidans* and its effect on metal partitioning. *Water Res.*, **36**(13): 3193-3202.

McCready, R.G.L. and LeGallais, B.P. 1984. Selective absorption of P-labelled *Thiobacillus ferrooxidans* in finely ground coal suspensions. *Hydrometallurgy,* **12**: 281-88.

Mercier, G., Chartier, M., Couillard, D. and Blais, J.F. 1999. Decontamination of Fly Ash and used lime from municipal waste incinerator using *Thiobacillus ferrooxidans. Environ. Manage.,* **24**(4): 517-528.

Miyamoto, S. and Enriquez, C. 1990. Comparative effects of chemical amendments on salt and sodium leaching. *Irrig. Sci.,* **11**: 83-92.

Pooley, F.D. 1982. Bacteria accumulate silver during leaching of sulphide ore minerals. *Nature,* **296**: 642-643.

Pooley, F.D. and Atkins, A.S. 1983. Desulfurization of coal using bacteria by both dump and process plant techniques. *See Ref.,* 132, pp. 511-26.

Rhoades, J.D., Kandiah, A. and Marshal, A.M. 1992. The use of saline water for crop production. Irrigation and drainage, paper 48, Rome, FAO, p. 135.

Robinson, P.C. 1983. Mineralogy and treatment of refractory gold from the Porgera deposit, Papua New Guinea. *Trans. Inst. Min. Metall. Sect. C: Miner. Process. Extr. Metall.,* **92**: C82-C89.

Rodriguez-leiva and Tributsch, M. 1988. Morphology of bacterial leaching patterns by *Thiobacillus ferrooxidans* on pyrite. *Arch. Microbial.,* **149**: 401-5.

Rojas, J., Giersig, N. and Tributsch, H. 1995. Sulfur colloids as temporary energy reservoirs for *Thiobacillus ferrooxidans* during pyrite oxidation. *Arch. Microbial.* **163**: 352-56, 385.

Rupela, O.P. and Tauro, P. 1973. Utilization of Thiobacillus to reclaim alkali soils. *Soil Biol. Biochem.,* **5**: 899-901.

Sakaguchi, H. and Silver, M. 1976. Microbiological leaching of a chalcopyrite concentrate by Thiobacillus ferrooxidans. *Biotechnol. Bioeng.,* **18**(8): 1091-1101.

Schippers, A. and Sand, W. 1999. Bacterial leaching of metal sulfides proceeds by two indirect mechanisms via thiosulfate or via polysulfides and sulfur. *Appl. Environ. Microbiol.,* **65**: 319-21.

Solisio, C., Lodi, A. and Veglio, F. 2002. Bioleaching of Zinc and aluminium from industrial waste sludges by means of *Thiobacillus ferrooxidans. Waste Manag.,* **22**(6): 667-675.

Stamford, N.P., Silva, A.J., Freitas, A.D. and Araujo Filbo, J.T. 2002. Effect of sulphur inoculated with *Thiobacillus* on soil salinity and growth of tropical tree legumes. *Bioresource. Technol.,* **81**(1): 53-59.

Sublette, K R., Kolhatkar, R. and Raterman, K. 1998. Technological aspects of the microbial treatment of sulphide-rich wastewaters: a case study. *Biodegradation,* **9**(3-4): 259-271.

Sugio, T., Domatsu, C., Munakata, O., Tano, T. and Imai, K. 1985. Role of ferric-ion-reducing system in sulphur oxidation of *Thiobacillus ferrooxidans. Appl. Environ. Microbiol.,* **49**: 1401-1406.

Torma, A.E. 1971. Microbiological oxidation of synthetic cobalt, nickel and zinc sulfides by *Thiobacillus ferrooxidans Rev. Can. Biol.,* **30**: 209-16.

Torma, A.E. 1973. Microbiological extraction of cobalt and nickel from sulphide ores and concentrates. *Can. Patent No.* 2, 169, 263.

Torma, A.E. and Guay, R. 1976. Effect of particle size on the biodegradation of a sphalerite concentrate. *Nat. Can.,* **103**: 133-38.

Torma, A.E. and Sakaguchi, H. 1978. Relation between the solubility product and the rate of metal sulfide oxidation by *Thiobacillus ferrooxidans. J. Ferment. Technol.,* **56**: 173-78.

Tributsch, 2001. Direct vs. indirect bioleaching. *Hydrometallurgy,* **59**: 177-85.

Visser, J.M., Stefess, G.C., Robertson, L.A. and Kuenen, J.G. 1997. *Thiobacillus* sp. Ws, the dominant autotroph oxidizing sulphide to sulphur in a reactor for aerobic treatment of sulphidic wastes. *Antonie Var Leeuwenhoak*, **72**(2): 127-134.

Yates, M.G. and Nason, A. 1966. Electron transport systems of the chemoautotroph *Thioobacillus ferrooxidans*. II. Purification and properties of a heat-labil iron cytochrome *c* reductase. *J. Biol. Chem.* **241**: 4872-4880.

Yudina, I.N., Kostyukova, O.N., Gorenkon, N.L., Panin, V.V. and Kim, D. Kh. 1979. Results of bacterial leaching on a scale-up continuous apparatus (vat process). *Tsvetn. Met.*, **8**: 109-111.

Zhu, N., Zhang, L., Li, C. and Cai, C. 2003. Recycling of spent nickel-cadmium batteries based on bioleaching process. *Waste Manag*, **23**(8): 703-8.

Soil Microbiology and Soil Biotechnology : Vol. 1
Editor : S. Kannaiyan
Associated Publishing Co., New Delhi, 2010, pp. 89-109

7

The Rhizosphere Plant Microbe Interaction

R. Mohan[1], L. Aruna[1], E. Subramanian[1], D. Rajakumar[2]
S. Sundaravarathan[1] and S. Kannaiyan[3]

Introduction

Plant physiologists have addressed plant nutrition exclusively for hydroponic conditions. Consequently, they have concentrated their research on the sole uptake processes. Similarly, most soil scientists have studied the dynamics of mineral nutrients in soils in the absence of plants. Though, these approaches have increased our knowledge of the mechanisms involved in nutrient dynamics, their major limitation has been to ignore or at least under estimate the importance of the interactions between plant roots and soil constituents for mineral nutrition. The chemical conditions in the rhizosphere can thus be drastically varied from those in bulk soil. A better prediction of nutrient bioavailability and mineral nutrition would thus take into account the chemical state and physical extent of the rhizosphere. These factors are either a direct effect of root activity itself or an indirect effect of the roots, *i.e.*, the effect of the root stimulated rhizosphere microflora.

With the notable exceptions of symbionts and pathogens, the study of soil biology in agriculture has historically dealt with the effect of agricultural practices on free-living organisms in the soil. Data from such studies are difficult to convert into integrated information that can be used to improve the productivity of crops and the viability of cropping systems. What is important agriculturally is how the interactions between management and soil biology affect the performance of crops. Roots are thus an integral component of the soil biology.

The concept of the rhizosphere was broached 100 years ago by the German biologist Lorenz Hiltner, who first proposed that the area around the roots is a region of high microbial activity. The term rhizosphere is derived from the Greek word "Rhizo or rhiza" means the root, the word "sphere" has many meanings, indeed we are dealing with a unique environment inhabited by a society of microorganisms, around a central point. The

1. Pandit Jawaharlal Nehru College of Agriculture & Research Institute, Karaikal - 609 603
2. Tamil Nadu Agricultural University, Coimbatore - 641 003
3. Chairman, National Biodiversity Authority, Chennai - 600 041

rhizosphere is that portion of the soil under the direct influence of the roots of higher plants with a different physical, chemical and biological environment from bulk soil. It is considered the more intense ecological habitat in soil in which micro organisms are in direct contact with plant roots. The root system of all higher plants is associated with a distinct, diverse community of metabolically active soil microbiota that carries out bio-chemical transformations.

Nature of rhizosphere and rhizodeposition

Extent of rhizosphere

A rhizosphere is born when a root tip arrives in a volume of soil and ends when that piece of root decays (Jones *et al.*, 2004). The rhizosphere of a single root occupies a volume that extends from the root to an ill-defined position in the soil that depends on the diffusion of exudates and the stage of development and biochemistry of the roots (Hinsinger *et al.*, 2005). The size of the rhizosphere may extend well beyond the first millimeters of soil adhering to plant roots when one considers the uptake of water and mobile nutrients (Darrah, 1998), however, the spatial extension of the rhizosphere varies widely for any single nutrient. The variations in the extension of the rhizosphere largely depends on the physical properties of soil that influence the transfer of ionic and molecular compounds and the resulting shape of their concentration gradients in the soil. Particle size, soil structure and water content among others are critical physical properties of soil that determine the geometry of the rhizosphere.

The spatial extension of the rhizosphere also varies with plant species. This partly relates to plant species varying in their strategies to further extend their rhizosphere through their development of root hairs and indirectly through mycorrhizal symbiosis (Bolon, 1991). Li *et al.*, (1991) showed that P which supposedly is taken up within several millimeters from the root surface can be acquired from a distance of up to 12 cm by the hyphae of mycorrhizal fungi. Plant species also differ in their rooting patterns. Some species develop original features such as proteoid roots, which result in large variations in the surface area of the soil – root interface and in the total volume of the rhizosphere soil used to acquire nutrients (Dinkelaker *et al.*, 1995).

The rhizosphere extent is variable, there being no sharp boundaries between it and the neighbouring soil. In the field, rhizospheres are much more complex. The roots of the plants are surrounded by a mucilaginous layer varying in composition from relatively simple oligosaccharides to a complex pectic acid polymer permeated by loose cellulose microfibrils. Electron micrographs give a graphic picture of this boundary zone between root and soil, indicating that the space between the cell wall and mineral soil particles is filled with gelatinous material known as 'mucigel'. The influence of plant roots extend beyond this mucilaginous zone, which may be 10-20 mm thick into the surrounding soil especially if the soil is not a fertile one.

Structure of rhizosphere

In the zone of root extension, the primary walls of epidermal cells are made up of cellulose microfibrils embedded in a thick, mucilaginous matrix of pectins and hemicelluloses. The root tip as it moves through the soil is protected by a root cap, the outer cells of which have mucilaginous walls which act as a lubricant. Behind the root cap is a meristematic region, 0.2-0.4 mm long, succeeded by a zone of elongation extending further 1-10 mm. Behind this again is a region of maturation or root hair zone up to several centimeters long where vascular tissues are differentiated (Fig. 1). During the phase of extension growth, epidermal cells and their associated root hairs release mucilages into the surrounding soil. Mucilaginous materials in the form of slime layers or capsules are also exuded or secreted by microorganisms. At the root – soil interface of older roots, plant and microbial exudates and secretions become mixed to produce the substance termed as '*mucigel*'.

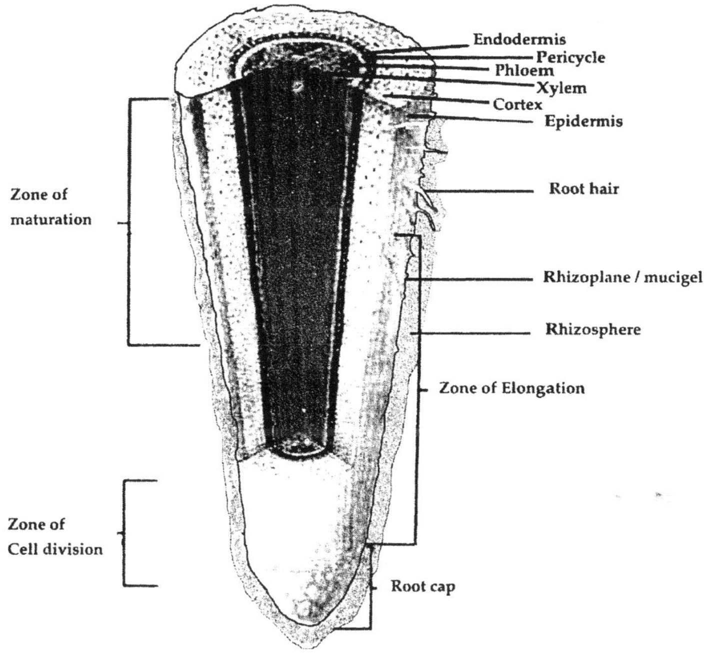

Fig. 1: Structure of rhizosphere with root zones

Although a rhizosphere effect can be discerned at a distance of several millimeters from the root surface in sandy soils and although the cylinder of rhizosphere soil which adheres to roots removed from soil is atleast a millimeter thick, the major zone of stimulation is measured in micrometers. There is also evidence of zonation within the rhizosphere. The part of the rhizosphere immediately adjacent to the root surface is termed as *rhizoplane* (0-1 mm). Further, it is differentiated into '*inner rhizosphere*" (upto 10 mm) and "*outer rhizosphere*" (10-20 mm). Similarly, bacteria may invade the empty cells of the epidermis and cortex behind the root hair zone and indeed the cortex of normal, healthy roots may be inhabited by a wide range of microorganisms. This specialized micro environment has been termed as '*endorhizosphere*'.

Organic substrates in the rhizosphere

In addition to the classical roles of providing mechanical support and allowing water/ nutrient uptake, roots also perform certain specialized roles, including the ability to synthesize, accumulate, and secrete a diverse array of compounds (Flores *et al.*, 1999). Rhizodeposition describes the total carbon transfer from plant roots to rhizosphere which comprises exudates, secretions, lysates and mucilage (Fig. 2). The ability to secrete a vast array of compounds into the rhizosphere is one of the most remarkable metabolic features of plant roots, with nearly 5% to 21% of all photosynthetically fixed carbon being transferred to the rhizosphere through root exudates (Marschner, 1995). Stimulation of microorganisms and chemistry of ion uptake is assumed to be caused by substances coming from roots, partly as a result of exudation and secretion and partly due to autolysis of moribund and dead root cells. Different plant species and also different parts of the root system of the same plant may have distinctive rhizosphere microflora and chemistry. The cause of these differences can be attributed to differences in rooting habit, tissue composition and in the product of exudation and secretion. The age of the plant also affects the composition of the rhizosphere flora. A rhizosphere effect is detectable in seedlings of few days old primarily due to the stimulatory effect of exudates/secretions. As the plant matures, dead and sloughed off tissues add appreciably to the rhizosphere effect. With the approach of senescence, it declines gradually and eventually becomes indistinguishable from the normal soil microflora.

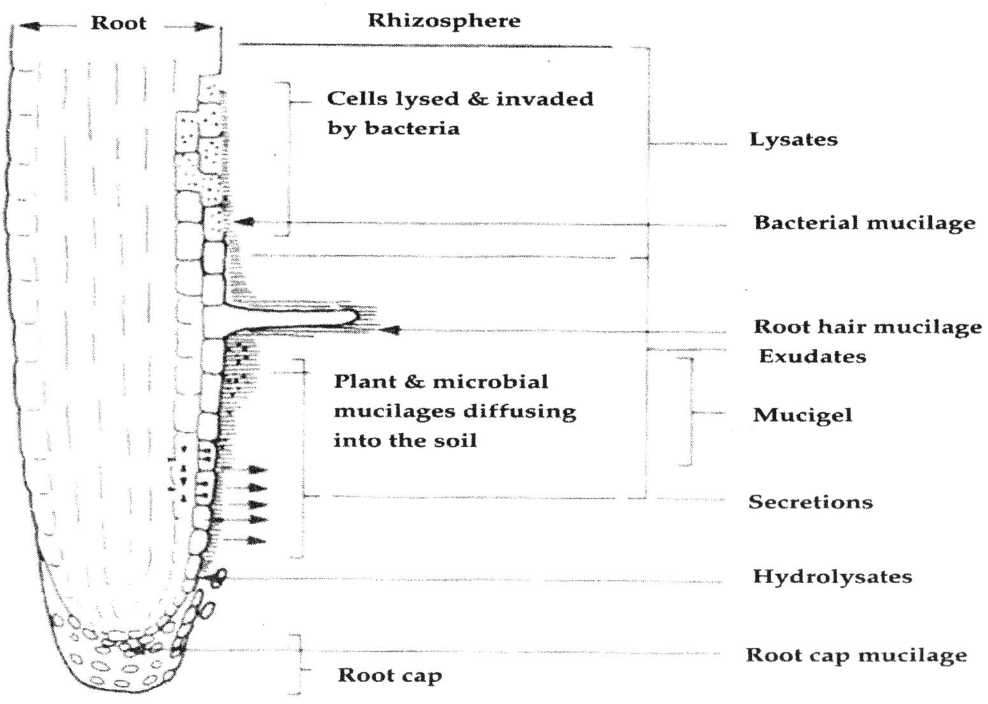

Fig. 2: Origin of organic substrates in the rhizosphere

Initially, microbiologists assumed that most of the organic substances escaped passively from roots independently of metabolic processes. It is now thought that some compounds are actively secreted as a result of the expenditure of metabolic energy. The root exudates include carbohydrates, amino acids, other organic acids, vitamins, nucleic acid derivatives and various other compounds (Table1). Among the carbohydrates exuded, at least ten sugars including an oligosaccharide have been identified in the wide range of plant species. The hexose's, glucose and fructose are generally most abundant. In contrast to their exudation of carbohydrates, plants appear to differ greatly in the amounts and kinds of amino acids they exude. The total amount of amino acid exudates from pea plant is many times that of oats.

Various other organic acids including several members of tricarboxylic acid cycle occur in root exudates which provide readily available substrate and important secondary effects such as alteration of pH and chelation of metal ions. Similarly, exoenzymes such as phosphatase and invertase, vitamins such as biotin and thiamine, miscellaneous compounds such as phenol derivatives are found in root exudates.

Table1: Organic compounds identified in the root exudates of various plants

Amino acids	Organic acids	Sugars	Vitamins	Purines/ nucleosides	Enzymes	Ions and gases
α-alanine	Citric	Glucose	Biotin	Adenine	Phosphatase	HCO_3^-
β-alanine	Oxalic	Fructose	Thiamin	Guanine	Invertase	OH-
Asparagines	Malic	Galactose	Niacin	Cytidine	Amylase	H+
Aspartate	Fumaric	Maltose	Pantothenate	Uridine	Protease	CO_2
Cystein	Succinic	Ribose	Riboflavin			H_2
Cystine	Acetic	Xylose				
Glutamate	Butyric	Rhamnose				
Glycine	Valeric	Arabinose				
Isoleucine	Glycolic	Raffinose				
Leucine	Piscidic	Deoxyribose				
Lysine	Formic	Oligosaccharides				
Methionine	Aconitic					
Serine	Lactic					
Threonine	Pyruvic					
Praline	Glutaric					
Valine	Malonic					
Tryptophan	Aldonic					
Ornithine	Erythronic					
Histidine	Tetronic					
Arginine						
Homoserine						
Phenylalanine						
γ-Aminobutyric acid						

(Dakora and Philips, 2002)

Factors affecting exudation

The principal factors that affect the kinds and quantity of substances released by roots into the rhizosphere soil include the species and development stage of plants, various soil physical – stress factors, plant nutrition, mechanical or disease injury, microbial activities and foliar – applied chemicals.

Numerous controlled experiments with plants grown in nutrient solutions or in sterile sand have provided convincing evidence that differences in the quantity and composition of exudates are likely to be greater among plants that are not closely related phylogenetically. The proportions of the various amino acids differed between peas and oats. Amino acid exudation by *Phalaris* and tomato exceeded that of clover growing in nutrient solution. Similarly, certain amino acids found in wheat were not in that of barley and *vice versa* (Rovira, 1959). While the nature of exudates varies between plant species, changes also occur with plant age or stage of development. The colonization of seeds and seedling roots by microorganisms depends to a large extent on the nutrients in the exudates provided by these sources. Thus, kinds and numbers of microorganisms change with plant growth as the chemical constituents of exudates change. A direct correlation can be found between seed size and amount of seed exudate released or the number of seeds needed to produce 1g of exudates (Vancura and Hanzlikova, 1972).

Though the rate of exudation generally increases for most plants with an increase of temperature above that which is optimal for plant growth, some plant species release greater quantities at low temperatures. Different species of plants respond differently to temperatures and therefore reveal different rates of photosynthesis and of translocation of photosynthates to the roots, different enzymatic activities that synthesize or degrade photosynthates, and variations in membrane permeability (Hale and Moore, 1979). Exudation of serine, glutamic acid, and á-alanine by clover roots in nutrient solution decreased when daylight intensity was reduced to 40% by shading (Rovira, 1959).

Stress created by low soil moisture in many cases increases exudation of various compounds and results in a higher microbial population density than under conditions of optimal moisture content (Katznelson *et al.,* 1955). It has been suggested that an increase in á-amino acids in plants as a result of proteolysis during wilting may contribute to an amino-acid imbalance resulting in an increased tendency to move outward (Kemble and Macpherson, 1954). It is evident; however, that soil moisture stress sufficient to interfere with water absorption and photosynthesis can be expected to induce changes in exudation patterns. In addition to the direct effect on the exudation process, soil moisture also influences the extent to which exudates can diffuse in the soil and affect microbial behavior. Gas exchange, which is closely linked with soil type and moisture, affects both root development and microbial activities.

Groundnut plant requires a relatively large amount of available calcium for peanut seed development and for this purpose gypsum is widely used in the cultural practices. Calcium is known to affect membrane permeability and changes in permeability may be related to the mechanism of root exudation (Vancura, 1967). In some cases nutrient factors may affect cell permeability, whereas in other cases cell metabolism may be affected, resulting in increased levels of compounds in the cytoplasm followed by increased loss from the roots.

Any foliar destruction of plants, such as clipping (defoliation), severe foliage disease, and herbicidal injury, interfering with the photosynthetic process can be expected to influence root exudation (Vancura and Stanek, 1975). Under normal agricultural practices, such as frequent clipping and grazing of forage crops, it can be expected that root exudates and microbial activity in the rhizosphere will fluctuate accordingly. Other forms of stress upon plants, not involving mechanical injury, can alter root exudation and the qualitative content of exudates. Salt-induced osmotic stress which occurs in high saline soils can be expected to affect membrane permeability, oxygen and nutrient uptake, and the movement of exudates from roots.

With higher concentration of carbohydrates, amino acids, vitamins and other growth-promoting substances, the rhizosphere is a zone of intense microbiological activity where the metabolic interaction between the rhizosphere microflora and green plants is extremely complex. The mucigel coating on the surface of roots provides a matrix in which microorganisms, soil nutrients, water, and gases are in intimate contact. Significant increases in root exudation often observed in the presence of microorganisms may result from an increased concentration gradient between root and nutrient solution created by microbial utilization of certain compounds in the exudates. Exudates from nodulated root systems are sometimes different from those in systems without *Rhizobium* nodules. Foliage-infecting microorganisms or viruses and root-infecting pathogens bring about changes in the metabolism of carbohydrates, amino acids, proteins, lipids, nucleic acids and natural growth regulators resulting in more root exudation and higher populations of microorganisms around roots (Hale and Moore, 1979). Increases in exudation induced by root pathogens can be attributed to direct effects in which root cells are injured and cell contents are released into the rhizosphere.

Root exudates can be altered by the application of some compounds to leaves. In few cases, herbicides have been shown to suppress exudation of principal compounds with the consequence of reduced microbial populations in the rhizosphere. A number of researchers have shown that foliar applications of plant nutrients may bring about changes in rhizosphere populations and this is attributed primarily to modification of root exudates. The concentration of amino acids generally increased and the sugar content in exudates decreased after treatment of foliage with nitrogen, whereas amino acids decreased and sugars increased following phosphorus treatment.

Rhizosphere in relation to soil characteristics and nutrient absorption

Soil characteristics

As a consequence of normal growth and development, a large range of organic and inorganic substances are secreted by roots into the soil, which inevitably leads to changes in its biochemical and physical properties (Rougier, 1981). Various functions have been attributed to root cap exudation including the maintenance of root-soil contact, lubrication of the root tip, protection of roots from desiccation, stabilization of soil micro-aggregates, and selective adsorption and storage of ions (Bengough and McKenzie, 1997 and Hawes *et al.*, 2000). Soil at field capacity typically possesses a matric potential of -5 to -10 kPa (Chaboud and Rougier, 1984). It has been speculated that as the soil dries and its hydraulic potential decreases, exudates will subsequently begin to lose water to soil. When this occurs, the surface tension of the exudates decreases and its viscosity increases. As the surface tension decreases, the ability of the exudates to wet the surrounding soil particles will become greater. In addition, as viscosity increases, the resistance to movement of soil particles in contact with exudates will increase, and a degree of stabilization within the rhizosphere will be achieved. For instance, McCully and Boyer (1997) reported that mucilage from the aerial nodal roots of maize has a water potential of -11 Mpa, indicating a large capacity for water storage when fully hydrated, whereas the mucilage loses water to the soil as it begins to dry.

This speculation supports the idea that root exudates could play a major role in the maintenance of root-soil contact, which is especially important to the plant under drought and drying conditions, when hydraulic continuity will be lost. The largest, most coherent soil *rhizosheaths* are formed on the roots of grasses in dry soil (Watt *et al.*, 1994). However, sheath formation requires fully hydrated exudates to permeate the surrounding soil particles that are then bonded to the root and each other as the mucilage dries. Young (1995) found that rhizosheath soil was significantly wetter than bulk soil and suggested that exudates within the rhizosheath increase the water-holding capacity of the soil. Modern cryo-scanning microscopy has helped researchers determine that the rhizosheath of a plant is more hydrated in the early morning hours compared with the midday samplings (McCully and Boyer, 1997). This implies that the exudates released from the roots at night allow the expansion of the roots into the surrounding soil. When transpiration resumes, the exudates begin to dry and adhere to the adjacent soil particles. Thus, the rhizosheath is a dynamic region, with cyclic fluctuations in hydration content controlled to some extent by roots. Taken together, these studies indicate that root exudation plays a major role in maintaining root-soil contact in the rhizosphere by modifying the physical properties of the rhizosphere and contributing to root growth and plant survival.

Nutrient absorption

The entry of ions into plant roots is accomplished by both passive and active mechanisms and the relative significance of these varies with soil conditions and plant species. The part of the root in which passive uptake processes such as diffusion and ion exchange operate consists of the cell wall and the inter cellular spaces. The cell wall surfaces due to the presence of carboxyl groups carry a negative charge and consequently are able to adsorb cations thereby freely accessible for diffusion. Early ideas about ion uptake from soils were based on the premise that CO_2 evolved by root respiration dissolved in water to form carbonic acid, H_2CO_3, which dissociates to provide H^+ ions to exchange adsorbed ions on soil colloids. In such a case, it is assumed that plants make the soil more acidic (by adding H^+). However, roots also have Cation exchange sites; hence the possibility arises for direct exchange between root and soil sites. In fact, the plants take more anions than cations consequently liberating HCO_3^- in order to balance the electro – chemical potential of the soil in the immediate vicinity of the roots.

Ionic concentration

The uptake of water and nutrients, which is the major function of plant roots results in either the accumulation or depletion of all the ions contained in the soil solution in the rhizosphere. This process occurs both for mineral nutrients and for other non essential elements. The nature and intensity of the changes in ionic concentrations depend on the correspondence of the requirements of the plant and the supply by the soil.

In calcareous soils, the accumulation of Ca can generate calcium carbonate precipitates around roots, a process that may impair plant growth through lime-induced chlorosis. Similarly, when Ca and sulphate accumulate concurrently in the rhizosphere, precipitation of calcium sulphate (gypsum) can occur. In saline soils, Na and Cl that occur in large concentrations in the soil solution can accumulate in the rhizosphere and reach much greater levels of concentration due to water intake by plant roots. Such accumulation of Na and Cl near the roots can severely impair plant growth and mineral nutrition, even for soils in which salinity would not be considered excessive due to its acceptable level in bulk soil conditions. In addition, the resultant increase in osmotic potential in the rhizosphere can restrict water uptake by roots, causing species that are not adapted to saline conditions to rapidly wilt and ultimately die.

The process of accumulating ions and salts in the rhizosphere are expected to be enhanced by an increased water uptake rate. Conversely, nutrients commonly occurring at low concentrations in the soil solution, such as K and even more so P, are transferred by the mass-flow process in amounts insufficient to meet the requirements of the plant. Their uptake thus results in a decrease in their concentration in the soil solution near plant roots; this depletion then generates a concentration gradient and diffusion of ions toward the roots.

A severe decrease in K concentration results not only in the diffusion of K toward the root but also in profound consequences for K dynamics. According to the mass-action law applied to ion exchange, it shifts the equilibrium of adsorption-desorption of K toward an enhanced desorption, leading to a depletion of exchangeable K and eventually to a depletion of non exchangeable K. This root induced release of non exchangeable K contributed upto 80 % of the uptake of the plants in soils where the release of non exchangeable K would have been expected to be negligible when considering the concentration of K in the bulk soil solution.

The depletion of P in the rhizosphere varies with different species. As in soil K, a severe decrease in soil P in the rhizosphere may cause a shift in the adsorption-desorption and dissolution-precipitation equilibria involved in the dynamics of soil P. Some of the studies suggest that P may sometimes accumulate in the rhizosphere rather than be depleted. It is concluded that the roots induced dissolution of hydroxyl apatite at a faster rate than that of P uptake, leading to the relative accumulation of P in the rhizosphere. The roots acted either through proton excretion or Ca uptake. For rye grass the proton excretion was likely to be the driving force of this root induced dissolution of phosphate rock occurring at a faster rate than that of P uptake. P can be expected to be depleted in the immediate vicinity of plant roots and to accumulate away from the roots. This would occur in the soil zone where dissolution due to proton excretion is faster than the diffusion of P toward the roots due to P uptake.

When considering the depletion of a nutrient in the rhizosphere, estimating its actual benefit for the plant is still difficult. Although nutrient depletion may result in the mobilization of less spatially or chemically accessible forms of the depleted nutrient as shown for K, it should result as well in a decrease in the uptake rate. This holds true unless concentration levels reached in the rhizosphere are maintained above the level required for maximum growth. Because of the lack of quantification of the different processes involved here, it is difficult to state if these phenomena may be regarded as efficient strategies of mineral nutrition. Undoubtedly, these root-induced changes of ionic concentrations deserve further consideration as major components of the whole process of nutrient acquisition by higher plants.

Rhizosphere pH

The uptake of nutrients can also affect soil pH. Indeed as nutrients are taken up as ions, the differential rates of uptake of cations and anions result in an imbalance of positive and negative charges entering the root cells. When an excess of cations over anions are taken up, the plant root compensates by releasing excess positive charges as protons, thereby resulting in acidification of the rhizosphere. Conversely, when an excess of anions over cations are taken up by the root, the excess negative charges are released as hydroxyls or bicarbonate ions, leading to alkalinization of the rhizosphere.

N is the most demanded mineral nutrient for numerous plants and since it can be taken up by the plant as an anion (as NO_3^-) as a cation (as NH_4^+), it has often been reported to play a major role in the overall cation-anion balance of the plant. Indeed, many studies have shown that plants supplied with ammonium acidify their rhizosphere, whereas they alkalinize it when supplied with nitrate. In addition, legumes relying on symbiotic N_2 fixation have been shown to acidify their rhizosphere to compensate for an excess of cations being taken up. This is however an oversimplified picture of the process, since, nitrate fed plants such as oilseed rape, chickpea or lupins have been repeatedly shown to excrete protons and acidify their rhizosphere. Some studies have also shown that the various parts of a single root exposed to identical, external conditions can behave differently, leading to localized processes of acidification and alkalinization.

Soil pH is known to be a critical factor influencing many chemical reactions in the soil environment. For instance, the dynamics of various forms of inorganic P are strongly pH-dependent, with dissolution of P from crystalline and sorption complexes and speciation of P in solution being strongly dependent on the pH of soil solution. In soils of moderate to high pH, some P reacts with Ca ions to form various sparingly soluble calcium phosphates such as octocalcium phosphate or hydroxyapatite, which require a supply of protons to dissolve and release P. Indeed, the ability of some species, such as buckwheat, oilseed rape and various legumes, to utilize P when supplied as a phosphate rock is related to their capacity to excrete protons. The proton excretion occurring especially when N is supplied as ammonium might improve P nutrition by enhancing the dissolution of some forms of inorganic P, most likely Ca-bound P in the rhizosphere. In flooded soils where adapted plants, such as lowland rice, are expected to rely solely on ammonium, root induced solubilization of acid-soluble soil phosphates has been shown to contribute a substantial proportion of P uptake. In addition, a stimulation of proton excretion has been reported for P-deficient species such as oilseed rape. In this respect proton excretion by plant roots may thus be regarded as an adaptive strategy for P acquisition.

The concentration of Fe in soil solution is severely decreased when pH increases, reaching a minimum for pH ranging from 7.4 to 8.5, due to the pH-dependent solubility of iron oxyhydroxides. A root induced decrease in rhizosphere pH would thus increase the activity of Fe in the soil solution by up to several orders of magnitude. In addition, many species have been shown to respond to Fe deficiency by acidifying their rhizosphere. Nevertheless, in the pH range commonly found in soils, Fe activity is always below 10^{-6} M, which is the value required for many plants to meet their Fe requirements. Such a value is attained only for soil pH of about 3. The extent to which proton excretion is capable of supplying a sufficient amount of Fe to roots for adequate plant growth is thus questionable unless very high fluxes of proton excretion occur at the soil-root interface.

Alkalinization of the rhizosphere is likely to be as widespread as acidification or even more so. Nitrate is the prominent source of N for non legume plants in most field conditions. The excretion of protons should not be regarded as a universal solution to the problems encountered by plants while acquiring nutrients. Rhizosphere acidification can have detrimental effects on root growth and mineral nutrition when soil pH is eventually decreased to very low values. The ryegrass fed with ammonium decreased its rhizosphere pH to 4.4 which is a pH value prone to aluminium toxicity. Thus it is clear that plant roots should not always acidify their rhizosphere and particularly not when growing in already acid soils so as to prevent increased risks of aluminium, manganese or even proton toxicity. Some plant species can increase rhizosphere pH in acid conditions and decrease it in neutral or alkaline conditions, revealing the capacity of plants to adapt to adverse soil conditions. Another major limitation of acid soils for plant growth is related to deficiencies in various mineral nutrients, especially P deficiency.

Whether root-mediated pH changes of the rhizosphere should be regarded as an adaptive strategy of nutrient acquisition or not, there is no doubt that the actual pH in the rhizosphere should be taken into account rather than the pH of the bulk soil when considering nutrient dynamics. Furthermore, the actual pH may be a poor indicator of the real effect exerted by the root on its environment. Considering that the activity of protons and hydroxyls is influenced by the total solute content of the rhizosphere solution and that a large part of protons or hydroxyl equivalents produced by the roots may be consumed in diverse adsorption-desorption or dissolution-precipitation reactions with soil minerals, the consequences of the excretion activity of the roots may extend much beyond what is indicated by a direct measurement of the resulting pH in the rhizosphere. The critical factor influencing the acquisition by plant roots of mineral nutrients such as P is thus the actual flux of protons of hydroxyl equivalents that are excreted at the soil-root interface rather than the resulting pH change.

Phytosiderophores and micronutrient availability

Some root exudates that act as metal chelators in the rhizosphere can increase the availability of metallic soil micronutrients, including iron, manganese, copper and zinc. Metal chelators form complexes with soil metals, thus releasing metals that are bound to soil particles and increasing metal solubility and mobility. The best evidence that plants use chelators in root exudates to increase micronutrient availability comes from research on the effects of graminoid phytosiderophores on iron (Fe) availability. Although Fe is often relatively abundant in soil, it is also often present as insoluble Fe (III) precipitates, particularly in soils with high or neutral pH. Graminoid secreted phytosiderophores bind to Fe (III) to form Fe (III) phytosiderophores, which grasses can take up with substantially greater efficiency than other chelated forms.

Phytosiderophores that have been identified include non proteinogenic amino acids such as mugenoic and avenic acid. Graminoid secretion of phytosiderophores is markedly greater in Fe-deficient that Fe-sufficient plants, indicating an important role of the compounds in mitigating Fe stress. Different grasses efficiently take up Fe (III) bound to phytosiderophores produced by other species, suggesting that phytosiderophore secretion may increase Fe availability across graminoid communities. The evidence that rhizosecreted chelators play a similarly important role in micronutrient availability to dicots is less strong than for graminoids.

Microbiology of rhizosphere

The rhizosphere effect of an actively growing crop plant is most pronounced with bacteria, followed by actinomycetes and fungi. Since the earliest review by Starkey (1929), the magnitude of root influence on microbial populations has been documented many times worldwide, revealing increased intensity of investigation up to recent times. It has been the practice to express the influence of roots on the microflora by means of the R/S ratio, which is the ratio of the microbial numbers per unit weight of rhizosphere soil (R) to the numbers in unit weight of adjacent non – rhizosphere soil (S). A better guide is provided by more recent work which aims at measuring growth rates on or around roots. Using this approach, it was reported that *pseudomonas* had a generation time of 5.2 h on roots of *Pinus radiata* compared to 77 h in unplanted soil while the corresponding values for bacilli were 39 h and more than 100 h, respectively. This not only shows the marked stimulatory effect of plant roots but also indicates the greater '*rhizosphere effect*' for *Pseudomonas* than for *Bacillus*.

Bacteria are the most abundant microorganisms in soil, whether determined by plate count or by direct microscopy. Counts recorded have varied between 10^6 and 10^8 cells per cubic centimeter of soil. Most estimates, however, are based on numbers per gram dry weight of soil and these have been estimated to be in excess of 100 million g^{-1} according to plate counts and more than 3 billion g^{-1} by microscopy procedures. Based on direct microscope estimates of populations at 8 million g^{-1} dry weight of soil, it has been estimated that bacteria make up between 0.03 and 0.28% of the total soil mass (Clark,1949).

Generally, gram negative bacteria are favoured over gram positive and non spore formers over spore formers. The short, gram negative rods that are greatly stimulated in the rhizosphere fall mainly into the three genera, namely *Pseudomonas*, *Achromobacter* and to a lesser extent, *Agrobacterium*. The high bacterial density in the rhizosphere undoubtedly results in a high degree of microbial competition and the selection pressures arising from this tend to favour rapidly growing and biochemically versatile organisms over slower growing and less versatile strains. This would suggest that the rhizosphere microflora has a greater ability to effect rapid biochemical changes than the general soil population.

The rhizosphere effect of actively growing crop plants usually corresponds to R/S ratio values of 2 to 20, but ratios of over 100 have also been recorded. The wide variation in the value is because of the different rhizosphere effects of different plant species, and it also varies with the stage of plant growth.

Most soil actinomycetes belong to three genera, *Streptomyces* and *Micromonospora* of the Streptomycetaceae and *Nocardia* of the Actinomycetaceae within the same taxonomic category as bacteria. Researchers often make reference to the Streptomycetes of soil or rhizosphere, since most species isolated fall within the class Streptomycetaceae, largely in the genus *Streptomyces*.

Populations of soil fungi in general fall in third place below those of bacteria and actinomycetes, based on dilution-plate counts. In contrast to their effect on bacteria, roots do not have such a marked influence on the total number of fungi. However, specific genera are stimulated, in other words the rhizosphere effect for fungi is more qualitative than quantitative. The imperfect fungi *Fusarium* and *Cylindrocarpon* are among the more prominent rhizosphere inhabitants but many other genera are represented, especially the *Zygomycetes*, *Mucor* and *Rhizopus*. Depending on the kind of plant and whether soil samples are from the outer rhizosphere, R/S ratios may range from 3 to over 200, but most frequently they are around 10 to 20 for crop plants.

Qualitative changes in rhizosphere microflora

Effects of higher plants on the qualitative nature of soil microorganisms may be more important than the stimulation of greater populations in general. Since plant taxa differ in chemical composition and physiology, and in the quantity and quality of root exudates, they can be expected to have some selective effects on specific microorganisms.

With the development of improved microbiological techniques, and extensive research on the subject, it is now well established that bacteria are more responsive to growth-stimulating factors from roots than are the actinomycetes, fungi, or algae. Both plate-culture and microscopy techniques have revealed that roots of most plants favor Gram-negative, non-sporing, rod-shaped bacteria, whereas Gram-positive, non-sporing rods and cocci, pleomorphic rods, and aerobic spore-forming bacteria are relatively less abundant directly adjacent to plants (Rovira and Davey, 1974). The rhizosphere harbours a higher percentage of motile bacteria and chromogenic types, as well as amylolytic, proteolytic, ammonifying, denitrifying, sugar-fermenting, and cellulose – decomposing bacteria, and those that can reduce methylene blue or resazurin. Most investigators agree that *Pseudomonas* spp. are among the most consistently abundant bacteria under the influence of roots. *Agrobacterium* and *Achromobacter* also are relatively common. While *Arthrobacter* has been found in large numbers in the rhizospheres of certain crops, species of this genus apparently are suppressed by the more competitive *Pseudomonas* (Katznelson, 1965).

Soil bacteria are regarded as relatively unstable physiologically, the incidence and activity of a particular group depending largely upon the availability of required nutrients. Therefore, it seems reasonable to expect that the physiological activities and nutrient requirements of bacteria in the rhizosphere, or at the root surface, may be somewhat different from those of bacteria in root-free soil. Within the rhizosphere the competitive interactions influencing the equilibrium of various groups depend on the availability of certain nutrients. Extensive work by Lochhead (1959) and Katznelson (1965) on the nutrition of soil and rhizosphere bacteria showed a higher percentage incidence in the root zone of bacteria requiring amino acids for maximal growth and lower relative numbers of those requiring nutrients supplied in soil extract.

The rhizosphere effect on nutritional groups of bacteria can be attributed both to substances released in root exudates and growth factors synthesized by other bacteria in the root zone. A higher percentage of bacteria capable of synthesizing growth factors required by other bacteria was found in the rhizoplane than in the rhizosphere of wheat. Over 50% of the rhizosphere and rhizoplane isolates tested were able to synthesize thymine, pantothenic acid, and riboflavin, while significant but smaller percentages produced other essential factors such as biotin and vitamin B_{12}.

Plant-microbe interactions

Plant-microbe interactions can positively influence plant growth through a variety of mechanisms, including fixation of atmospheric nitrogen by different classes of proteobacteria, increased biotic and abiotic stress tolerance imparted by the presence of endophytic microbes and direct and indirect advantages imparted by plant growth promoting rhizobacteria. Bacteria can also positively interact with plants by producing protective biofilms or antibiotics operating as biocontrols against potential pathogens or by degrading plant and microbe produced compounds in the soil that would otherwise be allelopathic or even autotoxic. In addition to providing a carbon-rich environment, plant roots initiate cross talk with soil microbes by producing signals that are recognized by the microbes, which in turn produce signals that initiate colonization. Chemical attraction of soil microbes to plant roots or chemotaxis, is a well understood mechanism involved in initiating cross talk between plant roots and microbes.

Rhizobium-legume symbiosis

Rhizobia belonging to the genera *Rhizobium*, *Sinorhizobium*, *Bradyrhizobium*, *Mesorhizobium* and *Azorhizobium* form symbiotic associations with leguminous plants by fixing atmospheric nitrogen in root nodules. However, rhizobial - legume interactions are very specific, allowing specific rhizobial strains to nodulate with specific host legumes. *Sinorhizobium meliloti* effectively nodulates species of the *Medicago, Melilotus* and *Trigonella* genera, whereas *Rhizobium leguminosarum* bv. *viciae* induces nodules in the *Pisum, Vicia, Lens* and *Lathyrus* genera. Conversely, not all members of the legume family form nodules. Of the 3 subfamilies of legumes, *Caesalpinoideae, Mimosoideae* and *Papilionoideae* members of the basal subfamily *Caesalpinoideae* are mainly non-nodulating. Thus nodulation and presumably nitrogen fixation are not ubiquitous within the legume family. The signal components largely responsible for these specific host microbe relationships belong to a class of compounds termed flavonoids. More than 4000 different flavonoids have been identified in vascular plants and a particular subset of them is involved in mediating host specificity in legumes. Isoflavonoids are only found members of the legume family. Daidzein and genistein, isoflavonoids produced by soybean effectively induce *Bradyrhizobium japonicum nod* genes.

Mycorrhizal associations

Mycorrhizal associations exist for prolonged periods with the maintenance of healthy physiological interactions between the plant and the fungus. The mycorrhizal associations of fungi and plant roots represent a diverse relationship in terms of both structure and physiological functions that leads to a nutrient exchange favourable to both partners. There are two basic types of mycorrhizal associations: Ectomycorrhizae and Endomycorrhizae. In Ectomycorrhizae, the fungal hyphae penetrate the intercellular spaces of the epidermis and of the cortical region of the root but do not invade the living cells. In contrast to the predominantly exogenous ectomycorrhizae, endomycorrhizae invade the living cells of the root, which become filled with mycelial clusters. Of the various microorganisms colonizing the rhizosphere, mycorrhizal fungi occupy the unique ecological position of being partly inside and partly outside the host.

Arbuscular mycorrhizal fungi (AMF) and plant roots form associations in more than 80 per cent of terrestrial plants. This symbiotic relationship increases nutrient uptake, improving plant fitness and in turn, the associated fungi extract lipids and carbohydrates from the host root. AMF may recognize the presence of a compatible host through root exudates, similar to recognition by rhizobia. The critical development step in the life cycle of mycorrhizal fungi is hyphal branching, which ensures contact with the host root and establishment of symbiosis. The branch inducing factor is a plant signaling molecule that triggers hyphal morphogenesis preceding successful root colonization. Root exudates from phosphate (P) - limited plants are more active than those from plants with sufficient P, suggesting that the production and/or exudation of branching factor in roots is regulated by P availability.

Plant growth-promoting rhizobacteria

The term rhizobacteria is used to describe bacteria that are able to colonize the root environment. Beneficial rhizobacteria otherwise called as Plant Growth Promoting Rhizobacteria (PGPR) are known to participate in many important ecosystem processes, such as the biological control of plant pathogens, nutrient cycling and/ or seedling growth through production of plant growth promoting substances. Bacteria are likely to locate plant roots through substances exuded from the root, and root exudates such as carbohydrates and amino acids stimulate PGPR chaemotaxis on root surface. In addition to fixing atmospheric nitrogen, *Azospirillum* spp. secrete phytohormones such as auxins, cytokinins and gibberellins. There is the exciting possibility that most PGPR are capable of producing growth regulators continuously, provided that precursors of phytohormones are available in the rhizosphere. Root exudates could supply the pool of precursors for rhizobacteria to biotransform. Moreover, it has recently been postulated that an additional mechanism for plant growth promotion by PGPR could be their altering of microbial rhizosphere communities (Ramos *et al.*, 2003)

Other rhizobacteria create "suppressive soils" by controlling plant diseases caused by soil fungi and bacteria. The mechanisms responsible for this biocontrol activity include competition for nutrients, niche exclusion, induced systemic resistance (ISR), and the production of antifungal metabolites. The bio control agents that are best characterized at the molecular level belong to the genus *Pseudomonas.* Most of the identified *Pseudomonas* bio control strains produce antifungal metabolites of which phenazines, pyrrolnitrin, 2, 4-diacetyl phloroglucinol (DAPG) and pyoluteorin are most frequently detected.

Microbe-microbe interactions

Direct interactions occurring between members of different microbial types often result in the promotion of key processes benefiting plant growth and health. Some PGPRs can improve nodulation and N_2-fixation in legume plants. Studies carried out under field conditions (Bai *et al.*, 2003), particularly those using ^{15}N-based techniques reinforce such beneficial co-operative effects between microbes. Research on the mechanisms by which PGPR enhance nodule formation implicates their production of plant hormones among the co-inoculation benefits. Chebotar *et al.*, (2001) demonstrated that some *Pseudomonas* strains, but not all, increased nodule number and acetylene reduction in soybean plants inoculated with *Bradyrhizobium japonicum.* Inoculation of phosphate-solubilizing bacteria (PSB) enhanced nodulation and N_2-fixation (^{15}N) by alfalfa plants, in parallel with an increase in the P content of plant tissues. It is therefore thought that an improvement in P nutrition of the plant resulting from the presence of PSB was responsible for increased nodulation and N_2-fixation, as it is well-known that these processes are P-dependent (Barea *et al.*, 2005).

Microbial populations in the rhizosphere are known either to interfere with or to benefit the establishment of mycorrhizal symbioses. A typical beneficial effect is that exerted by the 'mycorrhiza-helper-bacteria' (MHB), a term that was coined by Garbaye (1994) for those bacteria known to stimulate mycelial growth of mycorrhizal fungi and/or enhance mycorrhizal formation. This applies both to Ectomycorrhiza and to AM associations (Frey-Let *et al.*, 2005). Rhizosphere microorganisms are also known to affect the pre-symbiotic stages of AM development, such as spore germination rate and mycelial growth. The establishment of PGPR inoculates in the rhizosphere can be affected by AM fungal co-inoculation (Bianciotto and Bonfante, 2002). In particular, AM fungi inoculation improves the establishment of both inoculated and indigenous phosphate-solubilizing rhizobacteria acting as MHB (Barea *et al.*, 2002). Some AM fungi have established a particular type of symbiosis with endosymbiotic bacteria, previously assigned to the genus *Burkholderia* and recently reassigned to a new taxon named *Candidatus glomeribacter* (Jargeat *et al.*, 2004). These bacteria have interesting metabolic genes that may influence AM functions, and current investigations are aimed at exploiting this cooperative relationship in rhizosphere.

Bacteria can behave not only as individual cells but, under appropriate conditions where their numbers reach a critical level, they can modify their behaviour to act as multicellular entities. This is based on the fact that in natural ecosystems, bacteria do not exist as solitary cells, but are typically colonial organisms that live as consortia to exploit the elaborate system of intercellular communication which facilitates adaptation to changing environmental conditions. Microbial sensing and response mechanisms in the form of cell to cell communication via the use of small signalling molecules have recently been uncovered. Numerous cell density dependent signalling, molecule-mediated sensing and response pathways have now been defined, among which several fall within the scope of regulation that is commonly known as quorum-sensing (QS).

In the rhizosphere, several systems are known to have QS-regulated interactions, among which genera *Bradyrhizobium*, *Burkholderia*, *Erwinia*, *Photobacterium*, *Pseudomonas*, *Rhizobium* and *Sinorhizobium* have expression of several important genes under QS control. Within the PGPR, *Burkholderia cepacia*, *Pseudomonas chlororaphis*, *P. fluorescens*, *Rhizobium elti*, *R. leguminosarum*, *Sinorhizobium meliloti*, plant pathogens, *Agrobacterium rhizogenes*, *A. tumefaciens*, *Erwinia carotovora*, *E. chrysanthemi*, *E. stewartii* and *Pseudomonas syringae*, and saprophytes, *Chromobacter violaceum*, *Nitrosomonas europaea*, *Pseudomonas corrugata* and *Pseudomonas putida* exhibit QS system based on acyl homoserine lactones (AHLs) to communicate in the rhizosphere. Some strains of *B. subtilis* destroy the signal molecules of other systems in the rhizosphere and thus disturb the rhizosphere equilibrium. While current efforts are directed towards laboratory based assays of molecules involved in QS systems, their operation *in situ* in the rhizosphere appears imminent. Such information will permit delivery of not only more appropriate and effective bioinoculants for plant and soil health, but also cell-density-dependent control of *in situ* biological equilibrium in the rhizosphere.

The contribution of microbial co-operation in the rhizosphere to the formation and stabilization of soil aggregates has been demonstrated recently (Miller and Jastrow, 2000). Soil particles are bound together by bacterial products and by hyphae of saprophytic and AM fungi, into stable microaggregates. Microaggregates are then bound into macroaggregates with bacterial polysaccharides acting as binding agents and AM mycelia increasing the size of macroaggregates.

Molecular approaches to study rhizosphere microflora

Until recently, studies of plant–microbe interactions were based mainly on cultivation techniques. Such studies provided useful information regarding microbial diversity in the rhizosphere but suffered from bias, resulting from the media and cultivation condition applied, and from the inability to grow and isolate 99% of the natural microbial community (Prosser, 2002).

Broad-scale methods, such as phospholipid fatty acid analysis (PLFA), and DNA hybridisation and re-association kinetics, have provided culture-independent analyses of the microbial diversity in the rhizosphere. PLFA provides a broad-scale diversity index, such as the number of bacterial families present in the samples. Combinations of BIOLOG and PLFA techniques have shown differences in microbial community composition of bulk and rhizospheric soils. DNA re-association can also provide some information about the total genetic diversity of a given bacterial community (Torsvik and Ovreas, 2002). However, these methods cannot provide a higher resolution of microbial diversity where it is necessary to identify key microbial species at the community level or to elucidate their role in the ecosystem. These limitations have been overcome to some extent by using rRNA gene analysis for microbial diversity studies. Polymerase chain reaction (PCR) amplification of rRNA genes from environmental DNA samples, combined with fingerprinting techniques, such as denaturing gradient gel electrophoresis (DGGE), terminal restriction fragment analysis (T-RFLP), amplified rDNA restriction analysis (ARDRA), cloning and sequencing, provide detailed information about the species composition of whole communities.

Recent advances in molecular methods and genomics provide an exciting opportunity to redefine the relationship between plants and the microbes in their rhizosphere. Several techniques have been developed that allow microbial ecologists directly to correlate a specific metabolic activity with phylogenetically identifiable units in natural environments For example, the use of Fluorescence In Situ Hybridization (FISH) micro autoradiography could provide information about C-utilizing microbes and the presence of particular functional genes in these organisms. This approach could be further improved by the use of isotope microarrays that are not only able to detect more C-utilizing microbes but can also help to monitor the expression of key functional genes. The most exiting methods in the future will probably be combinations of Stable Isotope Probing (SIP) microarrays and SIP-metagenomics. SIP allows the isolation of entire C-utilizing genomes from environmental samples. The use of SIP in combination with microarrays will allow identification of microbes utilizing plant C exudates and estimations of the levels of expression of bacterial genes in the rhizosphere. Extracting heavy DNA from SIP experiments and using it to construct Bacterial Artificial Chromosome (BAC) libraries will help to determine structure function relationships of rhizosphere microorganisms. However, the success of these techniques depends on the refinement of the SIP technique that allows separation of heavy nucleic acid following pulse-labelling of the plant. Further, developments in functional genomics including proteomics and metabolomics will be useful to identify the genes expressed in the rhizosphere, while the use of promoters to drive gene expression specifically at the root – soil interfaces will allow the engineering of microorganisms for beneficial purposes.

Conclusion

There is great potential to use the wide genotypic and agronomically induced diversity of root systems and their exuded chemicals to influence rhizosphere biology to benefit crop production. Progress in the areas of pathogens and symbionts in this regard is clear. Further progress, especially related to interactions with nonpathogenic organisms, will rely on an appreciation of the properties of rhizospheres in the field: the spatial and temporal boundaries of these rhizospheres, and the effects of structural, chemical, and physical soil heterogeneity in which the roots and associated microorganisms exist and function.

The first agricultural revolution in the eighteenth century introduced crop rotation to take advantage of and manipulate microbial populations in the soil, although at that time it was not known why this benefited plant health and growth. The second revolution, which began in the 1960s and is described as the 'green revolution', was based on improved plant-breeding techniques and the development of hybrid varieties; it now includes genetic engineering of plants but also a heavy reliance on the use of chemicals. We may now be at the cusp of a third stage, which will combine both approaches in a more holistic and elegant strategy. Applying knowledge about beneficial plant-microbe interactions in the rhizosphere to plant-breeding and genetic-engineering technologies may allow us to increase food production while reducing stress on the environment and on global biodiversity.

However, most of the microbial biodiversity in soil remains unexplored and much work remains to be done to first identify and then characterize microorganisms that could be used in such applications. Furthermore, such approaches require a detailed knowledge of the molecular signalling that takes place between plants and microbes to drive expression of desirable traits and suppress unwanted effects in a controlled manner. Exploiting plants and microbes by using such an integrated approach requires a systems biology strategy to understand the degree and complexity of plant-microbe interactions.

References

Bai, Y.M., Zhou, X.M. and Smith, D.L. 2003. Enhanced soybean plant growth resulting from coinoculation of *Bacillus* strains with *Bradyrhizobium japonicum*. *Crop Sci.,* **43**: 1774-1781.

Barea, J.M., Toro, M., Orozco, M.O., Campos, E. and Azcon, R. 2002. The application of isotopic ^{32}P and ^{15}N-dilution techniques to evaluate the interactive effect of phosphate-solubilizing rhizobacteria, mycorrhizal fungi and *Rhizobium* to improve the agronomic efficiency of rock phosphate for legume crops. *Nutrient Cycling in Agroecosystems,* **63**: 35-42.

Barea, J.M., Werner, D., Azcon-Aguilar, C. and Azcon, R. 2005. Interactions of Arbuscular Mycorrhiza and nitrogen fixing symbiosis in sustainable agriculture. In: Agriculture, Forestry, Ecology and the Environment (eds.) Werner, D and W.E. Newton, Kluwer Academic Publishers, The Netherlands.

Bengough, A.G. and McKenzie, B.M. 1997. Sloughing of root cap cells decreases the frictional resistance to maize (*Zea mays* L.) root growth. *J. Exp. Bot.*, **48**: 885-893.

Bianciotto, V. and Bonfante, P. 2002. Arbuscular mycorrhizal fungi: a specialized niche for rhizospheric and endocellular bacteria. *Antonie van Leeuwenhoek Int. J. Gen. Mol. Microbiol.,* **81**: 365-371.

Bolon, N.S. 1991. A critical review on the role of mycorrhizal fungi in the uptake of Phosphorus by plants. *Plant Soil,* **134**: 189-207.

Chaboud, A. and Rougier, M. 1984. Identification and localization of sugar components of rice (*Oryza sativa* L.) root cap mucilage. *J. Plant Physiol.,* **116**: 323-330.

Chebotar, V.K., Asis, C.A. and Akao, S. 2001. Production of growth promoting substances and high colonization ability of rhizobacteria enhance the nitrogen fixation of soybean when inoculated with *Bradyrhizobium japonicum. Biol. Fertl. Soils,* **34**: 427-432.

Clark, F.E.1949. Soil microorganisms and plant roots. *Adv. Agron.* **1**: 241-288.

Dakora, F.D. and Phillips, D.A. 2002. Root exudates as mediators of mineral acquisition in low nutrient environments. *Plant Soil,* **245**: 35-47.

Darrah, P.R. 1998. Interactions between root exudates, mineral nutrition and plant growth. In: *Inherent variation in plant growth: Physiological mechanisms and ecological consequences.* (eds.)

Lambers, H., H. Poorter and M.M.I. Van Vuuren). Backhuys Publishers: The Netherlands. pp. 159-181.

Dinkelaker, B., Hengeler, C. and Marschner, H. 1995. Distribution and function of proteoid roots and other root clusters. *Bot. Acta*. **108**: 183-200.

Flores, H.E., Vivanco, J.M. and Loyola-Vargas, V.M .1999. "Radicle" biochemistry: the biology of root-specific metabolism. *Trends Plant Sci.,* **4**: 220-226.

Frey-Klett, P., Chavatte, M., Clausse, M.L., Courrier, S., Le Roux, C., Raaijmakers, J., Martinotti, M.G., Pierrat, J.C. and Garbaye, J. 2005. Ectomycorrhizal symbiosis affects functional diversity of rhizosphere fluorescent pseudomonads. *New Phytol.,* **165**: 317-328.

Garbaye, J. 1994. Helper bacteria: a new dimension to the mycorrhizal symbiosis. *New Phytol.*, **128**: 197-210.

Hale, M.G. and Moore, L.D. 1979. Factors affecting root exudation. **II**: 1970-1978. *Adv. Agron.* **31**: 93-126.

Hawes, M.C., Gunawardena, U., Miyasko, S. and Zhao, X. 2000. The role of root border cells in plant defence. *Trends Plant Sci.*, **5**: 128-133.

Hinsinger, P., Gobran, G.R., Gregory, P.J. and Wenzel, W.W. 2005. Rhizosphere geometry and heterogenity arising from root-mediated physical and chemical processes. *New Phytol.*, **168**: 293-303.

Jargeat, P., Cosseau, C., Olah, B., Jauneau, A., Bonfante, P., Batut, J. and Becard, G.. 2004. Isolation, free-living capacities and genome structure of '*Candidatus glomeribacter gigasporarum*', the endocellular bacterium of the mycorrhizal fungus *Gigaspora margarita. J. Bacteriol.*, **186**: 6876-6884.

Jones, D.L., Hodge, A. and Kuzyakov, Y. 2004. Plant and mycorrhizal regulation of rhizodeposition. *New Phytol.,* **163**: 459-480.

Katznelson, H. 1965. Nature and importance of the rhizosphere. In: Ecology of Soil - borne Plant Pathogens. (eds.) Baker, K.F. and W.C. Snyder, Univ. Calif. Press, Berkley. pp. 187-209.

Katznelson, H., Rouatt, J.W. and Payne, T.M.V. 1955. The liberation of amino acids and reducing compounds by plant roots. *Plant Soil,* **7**: 35-48.

Kemble, A.R and Macpherson, H.T. 1954. Liberation of amino acids in perennial rye grass during wilting. *Biochem J.,* **58**: 46-49.

Li, X.L., George, E. and Marschner, H. 1991. Extension of the Phosphorus depletion zone in VA – mycorrhizal white clover in a calcareous soil. *Plant Soil*, **136**: 41-48.

Lochhead, A.G. 1959. Rhizosphere microorganisms in relation to root – disease fungi. In: Plant Pathology Problems and Progress. (eds) Holton, C.S., G.W. Fischer, R.W. Fulton, H. Hart and S.E.A. McCallan. Univ. Wisc. Press, Madison, Wisconsin, pp. 327-338.

Marschner, H. 1995. Mineral Nutrition of Higher Plants (II ed), Academic Press, London.

McCully, M.E. and Boyer, J.S. 1997. The expansion of root cap mucilage during hydration: III. Changes in water potential and water content. *Plant Physiol*, **99**: 169-177.

Miller, R.M. and Jastrow, J.D. 2000. Mycorrhizal fungi influence soil structure. In: Arbuscular Mycorrhizas: Physiology and Functions. (eds) Kapulnik, Y and D.D. Douds Jr., Kluwer Academic Publishers, The Netherlands, pp. 3-18.

Prosser, J.I. 2002. Molecular and functional diversity in soil microorganisms. *Plant Soil*, **244**: 9-17.

Ramos, B., Garcia, J.A.L., Probanza, A., Barrientos, M.L. and Manero, F.J.G.. 2003. Alterations in the rhizobacterial community associated with European alder growth when inoculated with PGPR strain *Bacillus licheniformis. Environmental and Experimental Botany*, **49**: 61-68.

Rougier, M. 1981. Secretory activity at the root cap. In: Encyclopedia of Plant Physiology. (eds) W. Tanner and F.A. Loews. New Series Vol. 13B. Springler Verlag, Berlin, pp. 542-574.

Rovira, A.D. 1959. Root excretions in relation to the rhizosphere effect. IV. Influence of plant species, age of plant, light, temperature and calcium nutrition on exudation. *Plant Soil*, **11**: 53-64.

Rovira, A.D. and Davey, C.B. 1974. Biology of the rhizosphere. In: The Plant Root and its Environment (ed) Carson, E.W. Univ. Virgenia Press, Charlottesville, pp. 153-204.

Starkey, R.L. 1929. Some influences of the development of higher plants upon the micro organisms in the soil. I. Historical and Introductory. *Soil Sci*., **27**: 319-334.

Torsvik, V. and Ovreas, L. 2002. Microbial diversity and function in soil: from genes to ecosystems. *Curr. Opin. Microbiol*., **5**: 240-245.

Vancura, V. and Hanzlikova, A. 1972. Root exudates of plants IV. Differences in chemical composition of seed and seedling exudates. *Plant Soil*, **36**: 271-282.

Vancura, V. and Stanek, M. 1975. Root exudates of plants V. Kinetics of exudates from bean roots as related to the presence of reserve compounds in cotyledons. *Plant Soil*, **43**: 547-559.

Vancura, V. 1967. Root exudates of plants III. Effect of temperature and cold shock on the exudations of various compounds from seeds and seedlings of maize and cucumber. *Plant Soil*, **27**: 319-328.

Watt, M., McCully, M.E. and Canny, M.J. 1994. Formation and stabilization of rhizosheaths of *Zea mays* L. : Effect of soil water content. *Plant Physiol*., **106**: 179-186.

Young, I.M. 1995. Variation in moisture contents between bulk soil and the rhizosheath of *Triticum aestivum* L., *New Phytol*., **130**: 135-139.

Soil Microbiology and Soil Biotechnology : Vol. 1
Editor : S. Kannaiyan
Associated Publishing Co., New Delhi, 2010, pp. 110-128

8

Microbial Ecology of Arid Soils

A.V. Rao[1]

Plant arid and semi-arid ecosystems extend from the tropical to sub-alpine zones and from below sea level to above 3000 m altitude. Due to the extreme variations in geographic and climatic patterns, global arid environment consists of highly diversified and heterogeneous ecosystems. According to Meigs (1957), 19% (27.6×10^6 km^2) of the land area on the globe is covered by arid deserts and 14.6% (21.2×10^6 km^2) by semi-arid deserts. The expansion of deserts by human activities, often in combination with adverse climatic conditions, has necessitated for understanding the biological processes in extreme arid environments. The micro-flora of arid soils is basically similar to that of any other agricultural soils. But the harsh arid environment exerts a powerful selective force that is more readily observed with micro-organisms than with higher life forms because of short life span, high turn-over rates and high rates of spontaneous mutation. Further the microbial populations in arid soils are more dynamic because of higher temperatures and alternating wet and dry conditions. In this chapter an attempt was made to provide detailed accounts of micro organisms in arid soils in relation to its environment.

Distribution of micro-organisms

In general, micro-organisms play a significant role in various biochemical transformation in the soil and maintain soil fertility. The populations of different micro organisms in arid soils are generally found to be lower than in soils of other climatic zones (Meiklejohn, 1957). The northern Egyptian desert soils contained 0.08×10^6 to 0.81×10^6 bacteria per g soil and 0.7×10^4 to 9.8×10^4 actinomycetes per g soil (Abdel-Ghaffar and Fawaz, 1977). In Indian desert soils the popultion of bacteria varied from 0.022×10^6 to 0.89×10^6 per g soil while actinomycetes from 0.04×10^4 to 25.3×10^4 per g soil (Rao and Venkateswarlu, 1983). The populations of bacteria and actinomycetes in soils of various world deserts varied significantly and the variation might be due to differences in soil characteristics and climatic parameters.

Bacteria are the most numerous organisms in all the deserts followed by actinomycetes and fungi. Desert soils contain high proportion of actinomycetes to the total micro-organisms

1. Central Arid Zone Research Institute, Jodhpur - 342 003, India

than in any other soils (Ramaswami, 1966) which might be due to the greater resistance of this group to high soil temperature and desiccation. Meiklejohn (1957) reported 98% of the total microorganisms as actinomycetes in Kenyan desert soils. However, Hethner (1967) reported that actinomycetes represented only 2% of the population in the sandy soils of Tassili N' Ajjer. Among bacteria, Gram positive spore formers are the predominant group of bacteria found in Indian arid soils (Rao and Venkateswarlu, 1983). According to Mahmoud *et al.* (1964) Egyptian arid soils contained more than 32% spore formers as compared to 8% in cultivated agricultural soils. In Tunisian pre-Saharan desert soils, various Corynefoms constituted the dominant organisms in these soils followed by *Pseudomonas, Acinetobacter, Bacillus, Proteus* and *Micrococcus* (Skujins, 1977a).

The fungal population in different desert soils varied from 1×10^3 to as high as 261×10^3 (Killian and Feher, 1939; Lobova, 1960; Ranzoni, 1968). While in Indian Thar desert soils population of fungi varying from 0.2×10^3 to 8×10^3 was observed (Rao and Venkateswarlu, 1983). *Aspergillus* and *Penicillium* constituted nearly 80% of the total fungal population. But, *Cladosporium, Alternaria, Chaetomium, Fusarium, Helminsthosporium, Mucor* and *Cephalosporium* were also dominant. However, the generic and species composition of arid soil fungi is distinct for each desert. The myco-flora is dominated by fungi with dark coloured hyphae as those pigments act as solar light filters. Further hyphae of many xeric fungi are thick walled and constitute an additional light reducing device. Durrell and Shields (1960) reported that the survival time of *Stemphylium ilicis,* a characteristic fungus was 60 min. on irradiation at 253.7 nm, as against 2 min. for *Aspergillus niger.* An examination of temperature tolerance of fungi (Borut, 1960) conducted on several fungi isolated from Neger desert soils, showed that there are relatively few thermophilic fungi in desert soils. The main physiological function of the fungi are their ability to decompose organic matter rapidly and their contribution to better soil aggregate formation (Khudairi, 1969; Went and Stark, 1968).

Table 1: Distribution of various micro-organisms in arid soils (Rao and Venkateswarlu, 1983)

Organism	Population in different soils (g–1 dry soil)			
	Jaisalmer	Barmer	Jodhpur	Pali
Bacteria ($\times 10^5$)	0.22	0.75	12.10	32.67
Funyi ($\times 10^3$)	0.20	3.50	6.80	33.00
Actinomycetes ($\times 10^4$)	0.04	0.90	48.30	122.71
Azotobacter ($\times 10^3$)	0.01	0.28	3.90	5.40
Organic carbon (%)	0.069	0.155	0.179	0.593
Rainfall (mm)	90	120	370	400

There is a considerable variation in the populations of various micro-organisms in different arid soils (Table 1). In general, microbial populations in desert soils depend upon soil type, texture and organic matter content besides various climatic parameters such as rainfall, temperature etc. There was a positive correlation between organic matter content and the populations of different micro-organisms (r = 0.94 for bacteria, 0.85 for fungi, 0.27 for actinomycetes, 0.84 for *Azotobacter* and 0.80 for *Nitrosomonas* at the 95% confidence level), but numbers could not be correlated with rainfall and temperature. Thus the climatic parameters may be of only secondary importance while the organic matter content appeared to be directly influencing the micro-organisms.

In general, high temperatures and soil drying during summer months bring down the microbial populations to very low levels in many desert soils. (Sasson, 1972). Siefert (1970) concluded that the population of bacteria in a soil sample depends on moisture rather than on temperature. The change in the seasonal climatic conditions causes considerable variation in microbial numbers and activities. Rao and Venkateswarlu (1983) demonstrated a significant increase in the populations of different microorganisms during rainy period in Indian desert soil, but the lowest populations of different microorganisms was observed during summer months due to soil desiccation and the prevalence of high temperatures. The changes occurred not only with heterotrophic bacteria, but also with nitrifying bacteria. In Saharan soil, as the soil water content increased from 1.5% to 10%, the microbial numbers increased by 10 folds (Killian and Feher, 1935).

Effect of sand dune stabilization

Sand dunes are a dominant formation constituting 64.6% area of Western Rajasthan causing setback to agricultural lands, roads/rail tracks, buildings and settlements etc. Because of the high wind velocities in this region, the sand dunes move from one place to another. The technique of sand dune stabilization developed at this Institute consists of fencing of the area, establishment of micro wind breaks on the wind ward side in 5 m chess board pattern or in 5 m parallel strips and sowing of grass and transplanting the trees and shrub species with the onset of monsoon. Populations of bacteria, fungi, antinomycetes, nitrifying bacteria and *Azotobacter* bacteria were consistently higher in stabilised sand dunes than in unstabilised sand dunes. The difference was sometimes as high as 200 times (Venkateswarlu and Rao, 1981). Higher organic carbon and nitrogen, increasing C:N ratio nearing 10:1, lower soil temperatures and higher soil moisture content in stabilised dunes are the probable factors responsible for significantly higher activities of microorganisms (Aggarwal and Lahiri, 1977; Venkateswarlu and Rao, 1981). The variation in the populations of various microorganisms in different dunes might be due to degree of stablisation as well as variation in the annual rainfall and maximum temperatures attained.

Thermophilic organisms

Although many organisms in desert soils may survive surface soil temperatures reaching as high as 70°C, there are few thermophilic organisms present. With increase of the temperature, soil dries out rapidly, reaching soil water potential >40 M Pa, and no significant microbial proliferation is expected. However, Pochon *et al.* (1957) demonstrated that in Saharan soils there were 100-600 thermophilic bacteria per gram soil, and no nitrogen fixing and nitrifying thermophiles were present. Ammonification, denitrification, cellulolysis, hemicellulolysis and amylolysis were negligible at these high temperatures. As the temperature increased, the relative abundance of bacteria with simple nutrient need also increased, whereas bacteria requiring more complex nutrient media were favoured at lower temperatures (Elwan and Diab, 1970). However, there are no reports on the occurrence of thermophiles in Indian desert.

Effect of soil moisture

Thus, water appears to be a major climatic factor for microbial activities especially in desert soils, since soil water potential (Ψ), the force by which water is held in the microsites of soil, is related to the energy metabolism. To assimilate water held in soil at –0.1 M Pa *i.e.*, –1.0 bar, micro-organisms need approximately 2 Joules (0.5 calories) per mole of water (McLaren and Skujins, 1967). Therefore microbial numbers and their activities diminish greatly as the water content in soil decreases. Growth of different micro-organisms is restricted at different soil water potentials (Table 2). Fungi proliferate and initiate growth at lower water potential than that of the bacteria. Hence, fungi may be contributing more to decomposition of crop residues in dry soils.

Table 2: Approximate limiting soil water potentials for microbial growth at 25°C (Brown, 1976)

Water potential (M Pa)	Bacteria	Yeast	Fungi
–0.003	*Caulobacter*	–	–
–1.5	*Spirillum*	–	*Sclerotium rolfsii*
–3.0	Chemo autotrophs	–	–
–7.1	Most Gram-ve rods	Basidiomycetous yeasts	Basidiomycetes, *Rhizopus nigricans*
–15.0	Most cocci, *Lactobacillus, Bacillus*	Ascomycetous yeasts	*Fusarium, A. flavus,* Mucorales
–22.0	*Staphylococcus*	*Saccharomyces ruxiii Debaryomyces*	*Aspergillus niger, Paecilomyces variotii*
–31.0	–	–	*Penicillium*
–39.0	Halobacteria Halococci	–	*Aspergillus, Wallemia, Sporendonema sebi*
–49.0	–	–	*Chrysosporum, A. niger*
–59.0	–	–	*Erotium*
–70.0	–	–	*Xeromyces bisporus*

Dommergues (1962) while working with Western-North African soils observed that in the range of water potential from –1.5 to 10 M pa in the surface layer, the determinate plant species approached the senescent stage. But, micro-organisms were active thereby showing that micro-organisms can survive at lower water potentials than the plants. Further soil micro-organisms according to their response to water availability are classified into different groups. The first group referred as hyper-xerophilic micro-organisms includes the organisms which are involved in ammonification, amylolysis and decomposition of plant residues. The threshold value of water potential for this group is more than –8.0 M Pa. The second group xerophilic micro-organisms (–0.5 to –8.0 M Pa) participate in cellulose decomposition, nitrification and partially in amylolysis and ammonification. The third group comprises hydrophilic micro-organisms with a threshold value below –1.5 M Pa. These are carboxy methyl cellulose decomposers and di-nitrogen fixers. As the soil dries up, the activity of micro-organisms decreases and a qualitative modification of soil processes occurs as certain groups of organisms are rapidly inactivated compared to the others.

Effect of organic matter

Organic matter plays an important role in supporting higher populations of both flora and fauna. While studying the microbial populations of different Indian desert soils with varying organic carbon matter content, there was a positive and significant correlation between microbial population and organic matter content but not with temperatures and other climatic parameters. (Rao and Venkateswarlu, 1983). Similarly in response to the organic matter availability, the microbial activities are significantly higher under the plant canopies than in the plant free soil (Vollmer *et al.*, 1977; Wallace and Romney, 1972). With the addition of FYM in Indian desert soil, a significant enhancement in the populations of bacteria and fungi by 2.5 and 3.4 folds respectively was observed (Fig. 1).

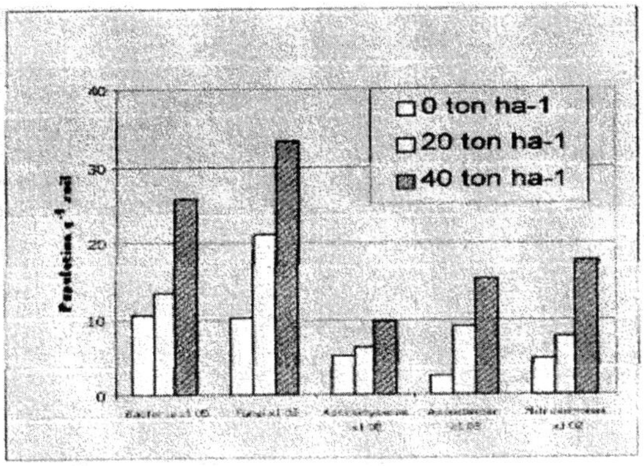

Fig. 1: Microbial population as influenced by organic matter content

There was no significant change in the composition of fungi with the addition of FYM, but the relative dominance of genera such as *Aspergillus, Penicillium, Fusarium* and *Mucor* increased markedly. Besides, nitrogen mineralising bacteria were also significantly enhanced indicating that organic matter is one of the limiting factors for optimum microbial activity in desert soils.

Biochemical activities

Biochemical processes occur at a greater rate in soils of arid climate than those of humid temperate climates, when water is not a limiting factor. The low organic matter content and the rapid rate of decomposition in aridsols is largely the result of higher temperatures. O'Brien (1978) observed that proteolysis in soils controlled primarily by environmental factors such as moisture and temperature rather than by the initial number or physiological characteristics of the native soil micro-organisms.

Skujins (1977a) demonstrated respiration, dehydrogenase activity, proteolytic activity in soils of different western deserts at water potentials from –7.7 to –2.8 M Pa. In Great Basin desert soils a positive correlation among dehydrogenase activity, respiration, proteolysis and nitrification was established (Skujins, 1973). The activity correlated well vertically with depths but correlation between sites was poor. Higher dehydrogenase activity was found in annuals and the activity was significantly higher in non-grazed areas compared to grazed areas indicating that the dehydrogenase activity might be used as a measure for the recovery process of grazed sites (Skujins, 1977b). Further dehydrogenase activity was influenced by the presence of plants. Tarafdar and Rao (1990) reported the dehydrogenase activity in Indian desert soils and varied with the crop. Application of organic manures significantly enhanced dehydrogenase activity compared to chemical fertilisers.

Table 3: Distribution of phosphatases in arid soil profiles under different vegetation (Tarafdar *et al.*, 1989)

Soil depth (cm)		Phosphatase activity*	
		Acid	Alkaline
	Crop land		
0-15		0.42	0.88
15-30		0.43	0.58
30-45		0.43	0.57
45-60		0.65	0.53
	Fallow land		
0-15		0.35	0.85
15-30		0.30	0.75
30-45		0.38	0.62
45-60		0.33	0.53
	Grass land		
0-15		0.32	0.67
15-30		0.30	0.48
30-45		0.37	0.50
45-60		0.62	0.43
	Tree land		
0-15		0.60	0.88
15-30		0.53	0.80
30-45		0.65	0.73
45-60		0.78	0.63
LSD (p = 0.05)		0.07	0.09

* expressed as enzyme units ($\times 10^{-2} g^{-1}$ soil)

Soil phosphatase plays an important role in converting organic phosphorus into plant available form especially in desert soils (Tarafdar *et al.,* 1989). Both plants and micro-organisms contribute towards building up of phosphatases in soils. Desert soils do possess acid and alkaline phosphatases and their activities are influened by the type of land use pattern. A significant increase in acid phosphatase activity was observed with depth beneath all types of vegetation while the activity was almost the same in all the layers in fallow land. On the other hand alkaline phosphatase was maximum in the surface layers compared to lower depths. (Table 3). Higher acid phosphatase activity in lower depths under different vegetation might be due to the secretion of the enzyme by plant roots which produce only acid phosphatase (Tarafdar and Claassen, 1988). Both phosphatases were significantly and positively correlated with soil organic matter, available and organic P, and soil microbial population.

Nitrifying bacteria

Population of nitrifying bacteria in various desert soils is low (Abdel-Salam *et al.*, 1971; Sasson, 1972; Rao and Venkateswarlu, 1983). The population varied from one soil to another within the same desert and there was a positive and significant correlation between rainfall, clay content, or organic matter with the population of nitriying bacteria (Abdel-Salam *et al.* 1971; Rao and Venkateswarlu, 1983). Stabilisation of sand dunes had enhanced the population of nitrifying bacteria by about 10 folds (Venkateswarlu and Rao, 1981). This was attributed to the favourable conditions caused by stabilisation of the soil ecosystem (Kaul, 1970; Aggarwal and Lahiri, 1977). With the availability of moisture in the soil especially during rainy season, the population of nitrifying bacteria is more compared to other seasons (Rao and Venkateswarlu, 1983). With the introduction of vegetation in arid soil, there is a significant improvement in nitrification process with the maximum activity in cultivated field followed by grasslands and woodlands, and with the least in fallows. With the application of organic manure, there is a rapid increase in the population of nitrifying bacteria. But there is a lag period in nitrification ranging from a few days to few weeks with the application of fertilisers (Alexander, 1978). Praveen-Kumar and Aggarwal (1988) observed 3 weeks as lag period for nitrification in arid soils while in normal soil it is about one week. In arid soils the population of nitrification was enhanced with the adoption of ley farming system (Rao *et al.*, 1977). Continuous pearlmillet cultivation supported maximum nitrifying bacteria followed by 3 years of mung bean or clusterbean indicating the dynamic changes in nitrification process with different crop rotations (Rao *et el.*, 1995). In deserts, the nitrifying population appears to be adopted to arid and warmer environments, and measurable nitrification may take place at water potentials below –1.5 M Pa (Skujins, 1984).

Nitrogen fixation

Biological nitrogen fixation in arid soils is mainly contributed by cryptogamic crusts, free living N_2-fixing bacteria, associative symbiotic nitrogen fixing bacteria and legume-*Rhizobium* symbiosis depending on the floral edaphic and climatic characteristics.The contribution of biological nitrogen fixers in arid zone is estimated to be 3.6 kg N ha^{-1} yr^{-1} as this process is restricted to a very small wet period.

Studies on N_2-fixation by cryptogamic crusts formed by blue green algae and lichens are restricted to temperate arid regions of North America (Klubek and Skujins, 1980; MacGregor and Johnson, 1971). Crust formation is a characteristic feature of the arid soils in shallow depressions during the wet periods. Nitrogen fixation by cryptogamic crusts on soil surface constitutes a major input of nitrogen to desert ecosystems (Rychert *et al.*, 1978). The average field N_2-fixation by surface cryptogamic crusts in Great Basin desert varied from 10-100 kg ha^{-1}. The highest rate of fixation occurs in the morning hours when the temperatures are relatively cool and the soil surface is moist because of dew. Further nitrogen fixation is inversely proportional to the water potential of the soil crust and N_2-fixation is negligible below –3.0 bars water tension. *Oscillatoria, Nostoc, Microcoleus, Nodularia, Scytonema, Tolypothrix* and *Schizothrix* are the common cyanobacteria associated with these crusts. Some of the cyanobacteria like *N. flagelliforme* from hot arid areas of China can survive drought for 2 years and yet on-rewatering, it takes up water very rapidly (Scherer *et al.*, 1984). It is able to recommence metabolic activity including nitrogen fixation during brief rainy periods. Blue green algae and/or lichens appear to be the major nitrogen fixers in the desert ecosystem with clay containing soils. (Table 4).

Table 4: Nitrogen fixation by cryptogamic crusts in different desert ecosystems (Rychert *et al.*, 1978)

Ecosystem	Ecosystem component	Estimated annual nitrogen input
Sonoran desert	Wet-dry cycling conditions of the crust	7-11 kg ha^{-1} yr^{-1}
Australian south central arid zone	Lichen (*Peltigera aphthosa*)	1.3 kg g^{-1} algae yr^{-1}
Sonoran desert	Pre-moistened algae crusts	13-18 kg ha^{-1} yr^{-1}
Great basin desert	Soil crusts	10-100 kg ha^{-1} yr^{-1}
South curfew valley validation site	Soil crusts *in-situ* measurements	7 kg ha^{-1} yr^{-1}

It has been observed that the desert soils do contain various algae belonging to chlorophyceae and cyanophyceae (Cameron and Fuller, 1960; Sankaran, 1980; Kaushik and Jain, 1989). These algal forms are not merely primary producers in the desert but they also possess potentialities for soil aggregation and promoting nitrogen fertility in desert soils besides checking soil erosion by crust formation (Sankaran, 1980).

Besides, algalization of sorghum and wheat for two crop seasons, each on saline sodic soils followed with saline water irrigation had decreased soluble Na/(Ca+Mg) ratio and exchangeable sodium in sandy clay loam calciorthid soils of arid Rajasthan (Jain and Kaushik, 1989), resulting in higher production of these crops and economising nitrogen application.

Free living nitrogen fixing bacteria

Among free living nitrogen fixing bacteria, *Azotobacter* has been extensively studied and was found to be ubiquitous in almost all soils including desert soils (Sasson, 1972; Pochon *et al.*, 1957). The desert soils contain very low populations and this is attributed to low available organic carbon, low water availability and high thermal regimes coupled with intense solar radiation (Vancura *et al.*, 1965; Venkateswarlu and Rao, 1981). Population varied from 0 to 11 × 10^4 cells per g in different soils of Indian arid zone and related to organic carbon content (Table 5). There is no doubt that growth and multiplication of this bacterium can be induced by enhancing the organic matter content as is evident from the higher populations in the rhizosphere soils of different crops or under different crop rotations (Rao *et al.,* 1995) or improvement in the soil moisture availability (Rao and Venkateswarlu, 1983). However micro-environmental conditions such as those occurring in the rhizosphere and within blue green algae-lichen crusts could provide a transient supply of available carbon resulting in higher nitrogen fixation by *Azotobacter* in arid soils. Using acetylene reduction technique it is estimated that non-symbiotic nitrogen fixation could account for 2 kg N ha^{-1} yr^{-1} and it may go up to 15 kg N ha^{-1} yr^{-1} depending upon the cropping pattern, organic matter status and soil moisture content (Rao and Venkateswarlu, 1982a) though Skujins (1984) considered nitrogen fixation by *Azotobacter* as a mechanism of nitrogen input in arid soils, a minor mechanism.

Table 5: Population of *Azotobacter* and organic matter content in different arid soils (Rao and Venkateswarlu, 1982)

Soil collected sites	Organic matter (%)	Population (× 10^2 g^{-1} soil)
Barmer	0.267	2.8
Bikaner	0.181	35.0
Churu	0.096	16.0
Dadaria	0.553	220.0
Degai	1.194	1100.0
Gowrisar	0.139	7.0
Jaisalmer	0.168	95.0
Jetpura	0.713	170.0
Jodhpur	0.309	39.0
Masitwala	0.184	160.0
Moulasar	0.501	28.1
Ramgarh (Unstabilised)	0.043	0.1
Ramgarh (Stabilised)	0.119	2.2
LSD (p = 0.05)	0.051	1.4

Associative symbiotic nitrogen fixation

Besides, symbiotic nitrogen fixation, the most popularly recognized input of nitrogen in the arid ecosystem is rhizospheric N_2-fixation associated mostly with non-legumes including succulent xerophytes (Rao and Venkateswarlu, 1982b, c). Since nodules were not obvious, these researches attributed the nitrogen fixation to *Azotobacter or Azospirillum* (Mahmoud *et al.,* 1964). All are heterotrophs depend on outside carbon source to fix nitrogen forming a loose or para symbiotic condition on the surface of the roots or growing in the outer layers of the root tissue as ecto-trophic or endo-trophic organizations. Bacterium isolated from the roots of tropical grasses of Indian arid zone and xerophytes was found to be *Azospirillum* and efficiency of these strains in respect to nitrogen fixation varied from one another. Most of these grasses belonging to C_4 type, and xerophytes (CAM type) have high level of water use efficiency and photosynthetic efficiency. (Steyn and Delwiche, 1970). Most of the strains of *A. brasilense* isolated from arid soils are tolerant to chlorides and sulphates, but are sensitive to chlorides (Rao and Tarafdar, 1990). This bacterium upon seed inoculation enhances dry matter production and grain yield of pearlmillet significantly (Joshi and Rao, 1989; Wani, 1990). An amount of nitrogen fertiliser equivalent to about 13-20 kg N ha^{-1} can be saved by employing *Azospirillum brasilence* as bio-inoculant (Joshi and Rao, 1989). Besides, production of arid grasses *viz. Cenchrus ciliaris* and *Lasiurus sindicus* can be enhanced by 10-12% by using *Azospirillum* (Rao and Venkateswarlu, 1988). Itzigsohn *et al.* (2000) indicated based on their field studies in Northern Negev desert, the potential of using PGPR like *Azospirillum brasilense* for growth promotion of natural pastures and further from ecological and environmental point of view, inoculation with *Azospirillum* contrarily to fertilization is relatively clean of negative impacts on the environment.

Symbiotic nitrogen fixation

The most popularly recognised input of nitrogen in any ecosystem is symbiotic nitrogen fixation through legume-*Rhizobium* association. As far as arid regions are concerned, the Caesalpinoideae are comparatively unimportant. Many of the genera belong to tropical rain forests. Some are found in arid and semi-arid areas for example, the apparently non-nodulating genus *Parkinsonia* dominates much of the Coloradan and Sonoran deserts (Felker, 1981). The Mimosoideae includes the very large genera *Acacia, Leucaena, Prosopis* and many other trees and shrubs together with some annual and herbaceous forms. These are extremely well adapted to arid climatic (Faria *et al.,* 1989). Nodulation is more or less universal and the nodules may be perennial. Felker and Clark (1982) had shown that nodulation and nitrogen fixation are dependent on the continuous presence of water and may occur at depths below 2.7 m. Normally nitrogen fixation is confined to wet periods and is related to shoot growth (Langkamp *et al.,* 1982; Rao, 1997). It is likely that perennials of the sub-family Papilionoideae may be important in most of the arid ecosystem, for example *Retama retam* and *Genista* spp. In the pre-Saharan desert extensive communities of Fabaceae shrubs exist in North American deserts such as *Cercidium* and *Prosopis.* However, the Papilionoideae contains nearly all the legumes currently exploited for grain and fodder purposes and grow under varied climates. A few are trees and the most are annuals. They normally nodulate and fix considerable quantities of nitrogen.

Attempts to locate nodules on *Prosopis* spp. in the field, usually up to 1 m depth, have given negative results and have caused speculation that these plants do not fix nitrogen in the field (Felker and Clark, 1982). Nodules in most of the arid legumes are often absent or very few. Beadle (1964), Beadle and Tchan (1955) and Rao (1989) attributed this due to lack of suitable conditions of temperature and moisture, and absence of proper endophyte. Since there are many legumes in the desert region, it is only reasonable to expect that nodulation of these plants occurs and is important in fixing nitrogen in such systems. A good cover of Swainsona (purple pea) may add up to 280 kg N ha^{-1} yr^{-1} to desert soils of Western New South Wales of Australia (Beadle, 1959). The plants often experience drought in arid climates because of low and erratic distribution of rainfall. In such dry soils, nodules may loose water faster than the vascular system can supply it and hence suffer water stress.

This may result in the collapse of cells near the surface giving impaired diffusion (Sprent, 1976). As nodules vary considerably in structure, they also show variation in their sensitivity to water stress (Table 6). Many of the nitrogen fixing trees produce perennial corky type nodules with suberised (corky) and in some cases lignified (woody) layers. These layers could serve as diffusion barriers preventing water loss and also oxygen uptake. This is consistent with a lack of N_2-ase activity in dry season. In wet periods such nodules develop aerating tissue. Clusterbean (*Cyamopsis tetragonoloba* (L.) Taub.) and moth bean (*Vigna aconitifolia* (Jacq.) Marechal) are the two important rainy season grain legumes growing extensively in Indian arid zone. As in other temperate/tropical legumes, these arid legumes also show genetic variability in nodulation and nitrogenase activity among different varieties (Khurana *et al.,* 1978; Rao *et al.*, 1984). These legumes have shown great variation in nodulation frequency, effectiveness of nitrogen fixation and response to inoculation and environmental stresses (Venkateswarlu *et al.,* 1982; Lynd and Ansman, 1982; Francois *et al.,* 1990; Arayangkool *et al.*, 1990). These legumes fix about 30-70 kg N ha^{-1} yr^{-1} depending on the rainfall, and a residual effect equivalent to about 15-20 kg ha^{-1} yr^{-1} is left over for the subsequent pearl millet crop. Inclusion of these arid legumes in crop rotation improves the soil biological productivity, resulting in higher pearl millet production (Rao *et al.,* 1995). In general arid legumes tolerate higher intensities of water stress because of their adaptation and their growth characteristics vis-à-vis high recovery capacity upon re-watering in these legumes (Rao and Venkateswarlu, 1987; Venkateswarlu *et al.,* 1983).

Table 6: Type of nodules and their sensitivity to water stress (Sprent, 1985)

Nodule type	Structure	Sensitivity to water stress
Determinate (spherical) soybean, moth bean, mung bean etc.	Limited growth period, no apical meristerm	Can withstand moderate water stress and then collapse/die with severe water stress
In-determinate (oblong) clusterbean	Longer growth period with apical meristem	Can withstand severe water stress. Growth can recommence after re-watering
Perennial corky (spherical) mainly shrubs and trees	Corky tissue grows in dry season in place of aerating tissue in wet season	Fixation only occurs in wet periods, may survive in inactive state for many months.

In arid ecosystem, legume trees play a significant role as these are important as a source of fodder, firewood, timber, fencing material as wind break/shelter belts etc. besides contributing nitrogen to the soil. Nitrogen fixing trees are widely grown either in agro-forestry or silvi-pastoral systems for sustainable production (Hocking, 1993; Tewari *et al.,* 2003). The ability of *Acacia, Prosopis, Albizia, Tecomella* etc. to tolerate droughts and adverse soil conditions especially salinity and alkalinity, and their successful adaptation to herbivore pressure, are the major reasons for their dominant position in arid zone (Fagg and Stewart, 1994). Nodulation in these trees is restricted to surface roots that too only during wet periods (Rao, 1997). Young trees exhibit better nodulation compared to old trees. When the nursery stock is healthy and sturdy, mortality rate is reduced and plants show better growth and development in out-plantations. Umali-Garcia *et al.*

(1981) and Sanginga *et al.* (1989) reported the improvement of stock of *Albizia lebbeck, Acacia mangium* and *Leucaena leucocephala* upon inoculation with different strains of *Rhizobium.* Amount of nitrogn fixed by these legumes vary from 36% to 54% (Rao and Giller, 1999; Sanginga *et al.,* 1989). Nitrogen fixed by the legume trees is not only utilized for growth but some amount is transferred to the non-legumes grown in between the trees. About 60-250 kg N ha^{-1} is being contributed to the soil nitrogen by these arid trees depending on the species and soil type (Pokhriyal and Mathani, 1929; Toky *et al.,* 1994).

Mycorrhizae

Most of the species have a highly specialized root fungus symbiotic relationship called mycorrhizae (Marx, 1972). The mycorrhizal symbiosis is an intimate association which exists between plant root systems and certain group of fungi. The relationship benefits plant growth by enabling available nutrients especially phosphorus, and water in the soil to be absorbed by the plants. Zueli He *et al.* (2002) indicated that the spatial fluctuations of the stressed soil environments in the Negev desert significantly influence the abundance and colonization of AMF. Spore density and colonization of AM structure groups have been shown to be sensitive indicators of spatial changes in AM-fungal distribution under the canopy of different desert plants. Arbuscular mycorrhizal (AM) hitherto referred as vesicular arbuscular mycorrhizal associations with plants are widely spread both taxonomically and geographically (Brundrett, 1991). The notable exceptions are members of the Amaranthaceae, Brassicaeae, Capaparaceae, Caryophyllacea, Chenopodiaceae, Commelinaceae, Cyperacea, Juncaceae, Portulacaceae, Proteaceae, Polygonaceae, Rosedaceae and Virticaceae (Kendrick and Berch, 1985). Out of 108 plant species (26 families) in the Indian arid zone, 86 were found to carry AM infections on their roots (Neeraj *et al.* 1991). While Mejstrik and Cudlin (1983) could not detect AM-fungal infections on the roots of 40 to 60% of the xerophytes belonging to different families. However, Kiran Bala *et al.* (1989) recorded mycorrhizal infections on all the trees of both legumes and non-legumes growing in Indian desert but per cent infection varies from 15 to 86% in different trees. In general, the young roots mostly carry mycelia extending throughout the length with a smaller number of vesicles, whereas spores and vesicles are found on mature roots. Besides, AM-fungal infections were found on the roots of various succulent xerophytes and the infection was significantly higher as compared to other vegetation (Kiran Bala, *et al.,* 1989). In contrast, Khan (1974) reported that none of the plants belonging to the family Euphorbiaceae was mycorrhizal. Mathew *et al.* (1991) observed that several species of cacti grow and flourish best in arid environments with root systems heavily mycorrhizal (Table 7). These AM-fungal associations may be considered as essential for the survival and growth of most of the plants including xerophytes growing in desert soils receiving a low amount of rainfall and especially in soils of low soil fertility (Lorgio *et al.,* 1999). However, these mycorrhizal associations can not be perceived as panacea for increasing plant production. But these associations appear to allow plants to extract nutrients and moisture from these desert soils for their survival. Similarly Rose (1981), Gerdemann (1961) and Miller (1979) reported the occurrence of these AM-fungal associations in different deserts. The presence of the AM-infections in desert vegetation helps in the stabilization of sand dunes through the binding of sand particles by the AMF mycelial threads as observed by Koske *et al.* (1975) in sand dunes of Lake Huron. The common genera of AM-fungi found in different arid soils are

Glomus, Gigaspora, Acaulospora, Sclerocystis etc. (Kiran Bala *et al.,* 1989; Mathew *et al.,* 1991). But recently Verma *et al.* (1998) have isolated a new root colonizing fungus from desert soils and identified as *Piriformospora indica gen. nov.* Further they were successful in cultivating this fungus in synthetic medium and said to have similarities to arbuscular mycorrhizal fungi (Varma *et al.* (1999).

Table 7: Mycorrhizal status of cactus roots growing in Indian desert (Mathew *et al.* 1991)

Host	Endophyte	Root infection (%)	Spores (x 50 g^{-1} soil)	
			Winter	Spring
Opuntia ficus-indica	*Glomus inferequence* *G. fasciculatum* *G. mosseae* *G. macrocarpum* *G. ferugianum* *Selecorystis sinuosa*	16	350	982
O. vulgaris	*G. reticulatum* *G. macrocarpum* *G. mosseae*	17	272	1142
O. cylindrica	*G. macrocarpum* *G. fasciculatum* *S. sinuosa*	32	251	384
O. santarita	*G. macrocarpum* *Gigaspora canodida*	31	328	443
Opuntia sp.	*Glomus geosporum* *G. macrocarpum* *G. feugianum* *G. fasciculatum*	12	244	332

It is well known that AM-fungi are not host specific. But the degree of AM-fungal colonization varies and its effect can differ with different host symbionts indicating thereby the host preference (Reena and Bagyraj, 1990b). Recent insights suggest that some degree of specificity exists between AM-fungi and host plants (Rosendahl *et al.,* 1992). Further, the functions of these AM-fungi vary from one species to the other. AM-fungal species of the genus *Gigaspora* appears to favour fluxes of carbon compounds from plant to soil biota resulting in enhanced soil aggregation while *Glomus spp.* tend to favor root colonization, plant growth and productivity (Varma, 1999). The degree of soil aggregation/stabilization vary due to differences in individual fungal species (Tisdall, 1991).

These fungi are not a stable group affecting plants only by their presence or absence but highly variable organisms that can effect a variety of plant responses. One such effect is a broadening of the ecological niche of plants through changes in the availability of nutrients. The potential host plant niche response mediated by AM-fungi is not only limited to enhanced nutrient uptake but may extend to the alleviation of the effects of excess minerals or presence of toxic metals (Allen, 1992; Arines *et al.,* 1989). Further Sylvia *et al.* (1993) and Newman *et al.* (1992)

had proposed that the AM-fungi help in transporting nutrients from higher fertility zones to lower fertility zones within the field through the spread of mycelial threads in the soil matrix. Thus mycorrhizae regulate the composition and functioning of plant communities in arid regions which are characterized by "Islands soil fertility". Possible benefits of AM-fungal inoculation include protection against root pathogens, an accelerated tolerance to environmental stresses, and perhaps the most important is reduction in use of artificial fertilizers besides improving the soil structure, thus higher plant productivity in arid soils (Neeraj *et al.,* 1991).

Molecular biology techniques in soil microbial ecology

The science of monitoring, survival, transport and ecological impact of genetically engineered micro-organisms in soil is rapidly advancing with improvements in sensitivity and specificity. These techniques are also instrumental in expending the knowledge of soil microbial ecology. Genetically engineered micro-organisms have the potential to mediate beneficial processes which could improve crop yields, degrade hazardous and toxic environmental pollutants, and perform leaching and recovery in mining and drilling operations. The study of microbial ecology, however have been advanced by the need to monitor the survival of GEMS. Many of the methods developed for detecting GEMs in the environment are being applied to the better understanding of the identity and role of the different members of the soil microbial community. Many studies of natural genetic exchange have been used as antibiotic resistant markers to track micro-organisms (Levy and Marshall, 1988). But these markers are not useful in soil where a number of indigenous micro-organisms are resistant to a wide range of antibiotics. So both genetic and phenotypic markers have been developed of detecting GEMs in soil (DeFlaun and Gerba, 1992).

Direct observation and quantification of specific micro-organisms in soil are possible by the fluorescent antibody technique based on immunological characters. Besides, *in-situ* detection of bacteria in soil has also been accomplished with introduced *E. coli* containing plasmids with genes for luminiscence (lux) originally cloned from *Vibrio fischeri.* Depending on the strain inoculated, as few as 2×10^{-2} cells g^{-1} of soil were detected by luminometry. Bioluminiscence detection systems such as charge coupled microscopy and x-ray film imaging, also have the potential to determine the spatial distribution of introduced organisms in soil (Rattray *et al.,* 1990).

Direct extraction of micro-organisms or DNA from soil and the use of nucleic acid probes to detect specific gene sequences within this 'communal genome'. Two different approaches are being used, one involving the direct extraction of total DNA from soil and the other is based upon separation of bacteria from soil followed by DNA extraction (Holben *et al.,* 1988).

Conclusion

Microbial ecology of arid soils had been studied extensively with respect to temperature variation, organic matter content, rainfall pattern etc. It is observed that arid soils contain a high proportion of actinomycetes and sporulating bacteria to the total microorganisms than in any other soils. Most of the microorganisms in desert soils are thermotolerant in nature. Different microorganisms in arid soils respond immediately to soil moisture and organic matter. Stabilization of sand dunes improved the microbial population by 200 folds with reference to that of un-stabilized dune soils. Similarly the activities of various enzymes, Vi., DHA and phosphatases were influenced by organic matter temperature, moisture etc. Besides, the population of nitrogen mineralizing bacteria (nitrifying and N2-fixing bacteria) were found to be very low. But these can be improved with organic matter content and soil moisture. Further arbuscular mycorrhizal fungi (AMF) are present in almost all arid soils with a large number of genera. The most common genera of AM-fungi are *Glomus* and *Gigaspora*. Possible exploitation of microbial diversity in arid soils for improving the plant productivity through use of various beneficial microorganisms such as nitrogen fixing bacteria (both free living and symbiotic types) and phosphate mobilising microorganisms including arbuscular mycorrhizal fungi have been discussed.

References

Abdel-Ghaffar, A.S. & Fawaz, K.M. 1977. Soil bacteria In: M.A. Ayyad (ed.), *Analysis of Mediterranean Desert Ecosystems of Northern Egypt,* Progress Report No. 3, pp. 1-13, University of Alexandria, Alexandria, Egypt.

Abdel-Salam, M.A., Monib, M., Abdel-Malek, Y. & Shady, M.A. 1971. Some factors governing the soil nitrifying capacity under the conditions of Kharga oasis soils. *A.R.E.J. Desert Institute,* **21**: 49-59.

Aggarwal, R.K. & Lahiri, A.N. 1977. Influence of vegetation on the status of organic carbon and nitrogen of desert soils. *Science and Culture*, **43**: 533-535.

Alexander, M. 1978. *Introduction to soil Microbiology*. Wiley Eastern India Limited, New Delhi.

Allen, M.F. 1992. *Mycorrhizal functioning. An integrative plant fungal process.* Chapman & Hall, New York.

Arayangkool, T., Schomberg, H.H. & Weaver, R.W. 1990. Nodulation and nitrogen fixation of guar at high root temperature. *Plant and Soil*, **26**: 209-213.

Arines, J., Vilarino, & Sainz, M. 1989. Effect of different inocula vesicular arbuscular mycorrhizal fungi on manganese content and concentration in red clover (*Trifolium pratense* L.) plants. *New Phytologist,* **112**: 215-219.

Beadle, N.C.W. 1959. Some aspects of ecological research in semi-arid Australia. In: Keast, A.T. (ed.) *Bio-geography and Ecology in Australia,* pp. 452-460. W. Junk, The Hague, Netherlands.

Beadle, N.C.W. 1964. Nitrogen economy in arid and semi-arid plant communities. III. The symbiotic nitrogen fixing organisms. *Proceedings of Linnean Society N.S.W.,* **89**: 273-286.

Beadle, N.C.W. & Tchan, Y.F. 1955. Nitrogen economy in semi-arid plant communities. I. The environment and the general consideration. *Proceedings of Linnean Society, N.S.W.,* **80**: 62-70.

DeFlaun, M.F. & Gerba, C.P. 1992. Monitoring Recombinant DNA Micro-organisms and Viruses in Soil. In: Metting Jr., F.B. (ed.) *Soil Microbial Ecology,* pp. 131-150, Marcel Dekker Inc., New York.

Francois, L.E., Donovan, T.J. & Mass, E.V. 1990. Salinity effects on emergence, vegetative growth and seed yield of guar. *Agronomy Journal,* **82**: 587-591.

Hethener, P. 1967. Activitie microbiologique des sols a *Eupressus dupreziana, A.* camus au Tassili N' Ajjer (Sahara Central). *Bull. Soc. Hist. Nat. Afr. Nord.,* **58**: 39-100.

Hocking, D. 1993. *Trees for drylands,* Oxford and IBH Publishing Co. Pvt. Ltd., New Delhi, p. 370.

Holben, W.E., Jansson, J.K., Chelm, B.K. & Tiedje, J.M. 1988. DNA probe method for the detection of specific micro-organisms in the soil bacterial community. *Applied Environmental Microbiology,* **55**: 3022-3025.

Itzogsohn, R.S., Burdman, S., Okon, Y., Zaady, E., Yovathan, R. & Perevolotsky, A. 2000. Plant growth promotion in natural pastures by inoculation with Azospirillum brasilense under suboptimal growth conditions. *Arid Soil Research and Rehabilitation,* **13**: 151-158.

Jain, B.L. & Kaushik, B.D. 1989. Effect of algalization on crop response under saline irrigation. *Journal of Indian Society of Soil Science,* **37**: 382-384.

Joshi, N.L. & Rao, A.V. 1989. Response of pearl millet to Azospirillum as influenced by N-fertiliser under field conditions. *Annals of Arid Zone,* **28**: 291-297.

Kaul, R.N. 1970. *The Afforestation in arid zone.* p. 155, Dr. W. Junk Publishers, The Hague.

Kaushik, B.D. & Jain, B.L. 1989. Algalization of saline soils and its influence on crop and soil properties. *Advances in Applied Phycology* II, pp. 43-49.

Kendrick, B. & Berch, S. 1985. Mycorrhizae applications in agriculture and forestry. In: Robinson, C.W. (ed.) *Comprehensive Bio-technology.* Vol. 4, Chap. 8, Pergamon Press, Oxford.

Khan, A.G. 1974. The occurrence of mycorrhizae in halophytes, hydrophytes and xerophytes and *Endogone* spores in adjacent soils. *Journal of General Microbiology,* **81**: 7-14.

Khudiari, A.K. 1969. Mycorrhiza in desert soils. *Biosciences,* **19**: 598-599.

Khurana, A.L., Saini, M.L., Jhorar, B.S. & Dudeja, S.S. 1978. Nodulation studies in clusterbean (*Cyamopsis tetragonoloba* (L.) taub.). *Forage Research,* **4**: 199-202.

Killian, C. & Feher, D. 1935. Recherches sur les phenomenes microbiologiques des sols sahariens. *Ann. Inst. Pasteur,* **55**: 573-623.

Killian, C. & Feher, D. 1939. Recherches sur la microbiologie des sols deserteques. *Encyl. Biol.* (Paris), **21**: 1-23.

Kiran Bala, Rao, A.V. & Tarafdar, J.C. 1989. Occurrence of VAM associations in different plant species of the Indian desert. *Arid Soil Research and Rehabilitation,* **3**: 391-396.

Klubek, B. & Skujins, J. 1980. Heterotrophic nitrogen fixation in arid soil crusts. *Soil Biology and Biochemistry,* **12**: 229-236.

Koske, R.E., Sutton, J.E. & Sheppard, B.R. 1975. Ecology and *Endogone* in Lake Huron sand dunes. *Canadian Journal of Botany,* **53**: 87-93.

Langkamp, P.J., Farnell, G.K. & Dalling, M.J. 1982. Nutrient cycling in a stand of *Acacia holoserirea A.* cunn. Ex G. Don. 1. Measurements of precipitation, interception, seasonal reduction, plant growth and nutrient requirement. *Australian Journal of Botany,* **30**: 87-106.

Levy, S.B. & Marshall, B.M. 1988. Genetic transfer in the natural environment. In: Sussman, M., Collins, C.H., Skinner, F.A. and Stewart-Tull, D.E. (eds.), *The Release of Genetically-Engineered Micro-organisms,* pp. 61-76, Academic Press, New York.

Lobova, E.V. 1967. *Soils of the Desert zone of the USSR.* Department of Commerce, Springfield, Virginia.

Lorgio, E.A., Julio, R.G. & Peter, L.M. 1999. Variation in soil micro-organisms and nutrients underneath and outside the canopy of *Adesimia bedwellii* (papilionaceae) shrubs in arid coastal Chile following drought and above average rainfall. *Journal of Arid Environments,* **42**: 61-70.

Lynd, J.Q. & Ansman, T.R. 1989. Soil fertility effects on nodulation, nitrogen fixation and yield of guar (*Cyamopsis tetragonoloba* (L.) Taub) grown on a typic Eutrustox. *Journal of Plant Nutrition,* **8**: 1069-1084.

MacGregor, A.N. & Johnson, D.E. 1971. Capacity of desert algal crusts to fix atmospheric nitrogen. *Soil Science Society American Proceedings,* **35**: 843-844.

Mahmoud, S.A.Z., Abou-El-Fadl, M. & El-Mofty, M.K. 1964. Studies on the rhizosphere microflora of a desert plant. *Folia Microbiology,* **9**: 1-8.

Marx, D.H. 1972. Mycorrhizae. In: McLaren, A.D. (ed.), *Agrichemical age.* pp. 13-16, Marcel Dekker, New York.

Mathew, J., Shankar, A., Neeraj & Verma, A. 1991. Glomaceous fungi associated with spineless cacti, a fodder supplements in deserts. *Transactions of Japanese Mycological Society,* **32**: 225-233.

McLaren, A.D. & Skujins, J. 1967. The physical environment of micro-organisms in soils. In: Gray, T.R.G. and Parkinson, D. (eds.), *The ecology of soil bacteria,* pp. 3-24. Liverpool University Press, Liverpool.

Meigs, P. 1957. Arid and semi-arid climatic types of the world. Proceedings of 17th Inter. Geogra. Cong. 1952, 8th General Congress Assembly, pp. 135-138.

Meiklejohn, J. 1957. Numbers of bacteria and actinomycetes in a Kenya Soil. *Journal of Soil Science,* **8**: 240-247.

Mejstrik, V.K. & Cudlin, P. 1983. Mycorrhiza in some desert plant species in Algeria. *Plant and Soil.* **71**: 363-366.

Miller, R.M. 1979. Some occurrence of vesicular arbuscular mycorhizae in natural and distributed ecosystems of the Red desert. *Can. J. Bot.,* **57**: 619-623.

Neeraj Shankar, A., Mathew, J. and Varma, A.K. 1991. Occurrence of VA-mycorrhizae within Indian semi-arid soils. *Biol. Fert. Soils,* **11**: 140-144.

Newman, E.I., Eason, W.R., Eissenstat, D.M. & Ramos, M.I.R.F. 1992. Interactions between plants: the role of mycorrhizae. *Mycorrhiza,* **1**: 47-53.

O'Brien, R.T. 1978. Proteolysis and ammonification in desert soils. In: West, N.E. and Skujins, J. (eds.) *Nitrogen in desert ecosystems,* pp. 50-59, Dowden Hutchinson and Ross, Stroudsburg, Pennsylvania.

Pochon, J., de Barjac, H. & Lajudie, J. 1957. Recherches sur la microflora des sols Sahariens. *Ann. Instt. Pasteur,* **92**: 833-836.

Pokhriyal, T.C. & Mathani, G.P. 1989. Techniques and approaches to utilize nitrogen fixing system in forestry plantation programmes. *Indian Forester,* **115**: 136-158.

Praveen-Kumar & Aggarwal, R.K. 1988. Reduction of ammonia volatilisation from urea by rapid nitrification. *Arid Soil Research and Rehabilitation,* **2**: 131-138.

Ramaswami, P.P. 1966. Studies on the physico-chemical and biological properties of soils of Madras. *Madras Agriculture Journal,* **53**: 388-387.

Ranzoni, F.V. 1968. Fungi isolated in culture from soils of the Sonoran Desert. *Mycologia,* **60**: 356-371.

Rao, A.V. 1989. Role of micro-organisms in plant nutrition under arid conditions. In: L.L. Somani, Bhandari, S.C. and Vyas, K.K. (eds.), *Biofertilizers,* pp. 67-80. Geo-environment Academia, Jodhpur.

Rao, A.V. 1997. Role of legumes along with biofertilizers in agroforestry systems for higher plant productivity. In: Gupta, J.P. and Sharma, B.M. (eds.), *Agroforestry for sustained productivity in arid regions,* pp. 89-98. Scientific Publishers, Jodhpur.

Rao, A.V. & Giller, K.E. 1993. Nitrogen fixation and its transfer from *Leucaena* to grass using N^{15}. *Forest Ecology and Management,* **61**: 221-227.

Rao, A.V., Kiran Bala & Lahiri, A.N. 1989. Influence of trees on micro-organisms of aridsol and its fertility. *Indian Forester,* **115**: 680-683.

Rao, A.V. & Tarafdar, J.C. 1990. Development of antibiotic resistant mutants of *Azospirillum brasilence* by UV-radiation, their performance on various salts and the growth of pearl millet. *Trans. ISDT,* **15**: 127-135.

Rao, A.V., Tarafdar, J.C., Sharma, S.K., Praveen-Kumar & Aggarwal, R.K. 1995. Influence of cropping systems on soil biochemical properties in an arid rainfed environment. *Journal of Arid Environment,* **31**: 237-244.

Rao, A.V., Singh, K.C. & Gupta, J.P. 1997. Ley farming—an alternate farming system for sustainability in Indian arid zone. *Arid Soil Research and Rehabilitation,* **11**: 201-210.

Rao, A.V. & Venkateswarlu, B. 1982a. Occurrence of *Azotobacter* and nitrogen fixation in the desertic soils. *Indian Journal of Microbiology,* **22**: 255-257.

Rao, A.V. & Venkateswarlu, B. 1982b. Nitrogen fixation by *Azospirillum* isolated from tropical grasses native to Indian desert. *Indian Journal of Experimental Biology,* **20**: 316-318.

Rao, A.V. & Venkateswarlu, B. 1982c. Associative symbiosis of *Azospirillum lipoferum* with dicotyledonous succulent plants of Indian desert. *Canadian Journal of Microbiology,* **28**: 778-782.

Rao, A.V. & Venkateswarlu, B. 1983. Microbial ecology of the soils of Indian desert. *Agricultural Ecosystems and Environment,* **10**: 361-369.

Rao, A.V. & Venkateswarlu, B. 1987. Nitrogen fixation as influenced by water stress in selected crop legumes of the Indian arid zone. *Arid Soil Research and Rehabilitation,* **1**; 89-96.

Rao, A.V. & Venkateswarlu, B. 1988. Influence of *Azospirillum* on the productivity of grasses of Indian desert. pp. 228-230. Proceedings of the 3rd Inter. Rangeland Congress, Rangeland Management Society of India, Jhansi.

Rao, A.V., Venkateswarlu, B. & Henry, A. 1984. Genetic variation in nodulation and nitrogenase activity of in guar and moth. *Indian Journal of Genetics and Plant Breeding,* **44**: 54-57.

Rattray, E.A.S., Prosser, J.I., Killham, K. & Glover, L.A. 1990. Luminiscence based non-extractive technique for *in-situ* detection of *Escherichia coli* in soil. *Applied Environmental Microbiology,* **56**: 3368-3374.

Rose, S.L. 1981. Vesicular-arbuscular endomycorrhizal associations of some desert plants of Baja, California. *Can. J. Bot.,* **59**: 1056-1060.

Rosendahl, S., Rosendahl, C.N. and Sochting, U. 1992. Distribution of VA-mycorrhizal endophytes amongst plants from a Danish grass land community. *Agric. Ecosyst. Environ.,* **29**: 329-335.

Rychert, R.C., Skujins, J., Sorensen, D. & Porcella, D. 1978. Nitrogen fixation by lichens and free living micro-organisms in deserts. In: West, N.E. and Skujins, J. (eds.), *Nitrogen in Desert Ecosystems,* pp. 20-30. Dowden. Hutchinson and Ross, Stroudsburg, Pennsylvania.

Sasson, A. 1972. Microbial life in arid environments: Prospects and achievements. *Annals of Arid Zone,* **2**: 67-91.

Sankaran, A. 1980. On some algae from the desert soils of Suratgarh, Rajasthan. *Annals of Arid Zone,* **19**: 389-394.

Sanginga, N., Mulongoy, K. & Ayanaba, A. 1989. Nitrogen fixation of field inoculated *Leucaena leucocephala* (Kam.) de Wit. estimated by the N15 and the difference methods. *Plant and Soil,* **117**: 269-274.

Scherer, S., Ernest, A., Chen, T.W. & Boger, P. 1984. Re-wetting of drought resistant blue green algae: time course of water uptake and re-appearance of respiration, photosynthesis and nitrogen fixation. *Oecologia,* **62**: 418-423.

Seifert, J. 1970. The influence of temperature on the numbers of bacteria in soil. *Acta Univ. Carol. Biol.,* **6**: 471-473.

Skujins, J. 1973. Dehydrogenase—An indicator of biological activities in arid soils. *Bulletin of Ecological Research and Communication (Stockholm),* **17**: 235-241.

Skujins, J. 1977a. Soil microbiological and biochemical investigations, 1972-1976. In: G. Novikoff, G., Wagner, F.H. and Hajjaj, M.S. (eds.): *Systems Analysis of pre-Saharan Ecosystem of Southern Tunisia.* Report No. 6, pp. 215-249. Utah State University, Logan.

Skujins, J. 1977b. *Comparison of biological processes in Western deserts.* US/IBP Desert Biome Res. Memo 75-33, Utah State University, Logan.

Skujins, J. 1984. Microbial ecology of desert soils. In: Marshall, K.C. (ed.) *Advances in Microbial Ecology,* pp. 49-91. Plenum Press, New York.

Sprent, J.I. 1976. Water deficits and nitrogen fixing root nodules. In: Kozlowski, T.T. (ed.) *Water deficits and plant growth, vol. 4, Soil water measurement, plant responses and breeding for drought resistance,* pp. 291-316., New York, Academic Press.

Sprent, J.I. 1985. Nitrogen fixation in arid environments. In: G.E. Weickens, G.E., Goodin, J.R. and Field, D.V. (eds.) *Plants for arid lands,* pp. 215-230, George Allen and UnWin, London.

Steyn, P.L. & Delwiche. C.C. 1970. Nitrogen fixation by non-symbiotic micro-organisms in some Californian soils. *Environmental Science and Technology,* **4**: 1122-1128.

Sylvia, D.M., Hammond, J.M., Bennett, J.H., Hass & Linda, S.B. 1993. Field response of corn to water management and inoculation with VAM. *Journal of Arid Environments,* **4**; 203-207.

Tewari, J.C., Harsh, L.N., Pratap Narain, Tripathi, D., Bohra, M.D. & Kaushish, S. 2003. Hardwickia binata thrives in the Indian Thar desert. *Asia-Pacific Agroforestry Newsletter,* No. 21, pp. 6-7.

Tisdall, J.M. 1991. Fungal hyphae and structural stability of soil. *Aust. J. Soil Res.,* **29**: 729-743.

Venkateswarlu, B., Rao, A.V. & Lahiri, A.N. 1983. Effect of water stress on nodulation and nitrogenase activity of guar (*Cyamopsis tetragonoloba* (L.) Taub.). *Proceedings of Indian Academy of Sciences (Plant Sciences),* **92**: 297-301.

Varma, A. 1999. Functions and applications of arbuscular mycorrhizal fungi in arid and semi-arid soils. In: *Mycorrhiza* (A. Varma and B. Hock eds). Springer Verlag, Berlin, pp. 521-556.

Verma, S., Varma, A., Rexer Karl-Heinz, Hassel, A., Kost, G., Sarbhoy, A., Bisen, P., Buetehorn, P. & Franken, P. 1998. *Piriformospora indica* Gen. Nov., a new root colonizing fungus. *Mycologia,* **90**: 895-909.

Vollmer, A.T., Au, F. & Bamberg, S.A. 1977. Observations on the distribution of microorganisms in desert soil. *Great Basin Nat.,* **37**: 81-86.

Wallace, A. & Romney, E.M. 1972. In: Wallace, A. and Romney, E.M. (ed.), *Radio ecology and ecophysiology of desert plants in the Nevada test site.* pp. 352-357. US Atomic Energy Commission.

Wani, S.P. 1990. Significance of *Azospirillum* on pearl millet production in semi-arid conditions. *Indian Journal of Microbiology,* **30**: 363-393.

Went & Stark, 1968. The biological and mechanical role of soil fungi. *Proceedings of National Academy of Sciences,* USA, **60**: 497-509.

Xueli, He, Mourato, S. & Steinberger, Y. 2002. Spatial distribution and colonisation of arbuscular mycorrhizal fungi under the canopies of desert halophytes. *Arid Soil Research and Rehabilitation,* **16**: 149-160.

Soil Microbiology and Soil Biotechnology : Vol. 1
Editor : S. Kannaiyan
Associated Publishing Co., New Delhi, 2010, pp. 129-135

9

Soil Microbial Diversity from the Himalaya: Need for Documentation and Conservation

*Anita Pandey, Pankaj Trivedi, Bhavesh Kumar, Bhaskar Chaurasia and LMS Palni**

Introduction

The Himalayan region represents great variation with respect to altitude, latitude, longitude, topography, geology, precipitation and edaphic conditions, etc.; this in turn is responsible for the occurrence of a wide variety of habitats and climates. It provides varied ecological niches and microclimates not only for the higher plants and animals but for the microorganisms as well. Systematic work was initiated in this laboratory on the isolation, identification and characterization of microbial communities of the region about 15 years ago. While the temperate and alpine locations have been explored for enumerating the diversity of free-living bacterial, actinomycetes and fungal communities, the symbiotic associations between selected trees and the AM (arbuscular mycorrhizal) fungi have also been investigated. Some work has also been carried out on microbial diversity of two hot spring sites, located in the Garhwal Himalaya, for the exploration of thermophiles.

Bacterial, actinomycetes and fungal diversity in rhizosphere and non rhizosphere soils

Bacteria, actinomycetes and fungi are three major groups of soil inhabiting microorganisms. Diverse vegetation, including trees, harbour a variety of specific rhizospheres allowing colonization by selected groups of soil microorganisms. While edaphic as well as climatic conditions affect the number and nature of microbial diversity in general, factors like root exudates and age of the host plants affect the microflora associated with a given rhizosphere, in addition. In a case study, conducted for the enumeration of microbial communities along an altitudinal gradient in the Mamlay watershed, Sikkim Himalaya, generally the microbial counts were found to decrease with increasing altitude (upto 1200-1900 m amsl). On the other hand, certain groups of microorganisms, such as Pseudomonads

Environmental Physiology and Biotechnology, G.B. Pant Institute of Himalayan Environment and Development, Kosi-Katarmal, Almora, 263 643.

* (Present address) State Biotechnology Programme, Govt. of Uttaranchal, Biotech Bhavan, P.O. Haldi, Pantnagar, U.S. Nagar 263 146.

and some other pigmented bacteria, showed an increase in the microbial counts with increasing altitude. However, one of the dominant bacterial species of the genus *Bacillus* did not show much variation in terms of counts, probably on account of its endospore forming ability. Amongst fungi, species of *Penicillium, e.g., P. janthinellum, P. javanicum, P. pinophillum*, and *P. raistrickii* were found to dominate the soil microflora at higher altitudes. Species of *Aspergillus, Cladosporium, Epicoccum, Fusarium, Gangronella, Myrothecium, Paecilomyces* and *Trichoderma* also contributed to the fungal flora. Various actinomycetes, mainly those that produce diffusible pigmented metabolites, were also isolated from these soils (Pandey and Palni, 1998a,b). Similar observations were recorded in investigations relating to the microbial diversity in Uttaranchal Himalaya; soil samples in these studies were collected from 1200 m to 3600 m (unpublished observations). Representative examples of selected and frequently occurring species of bacteria, actinomycetes and fungi have been shown in Fig.1 (A-F).

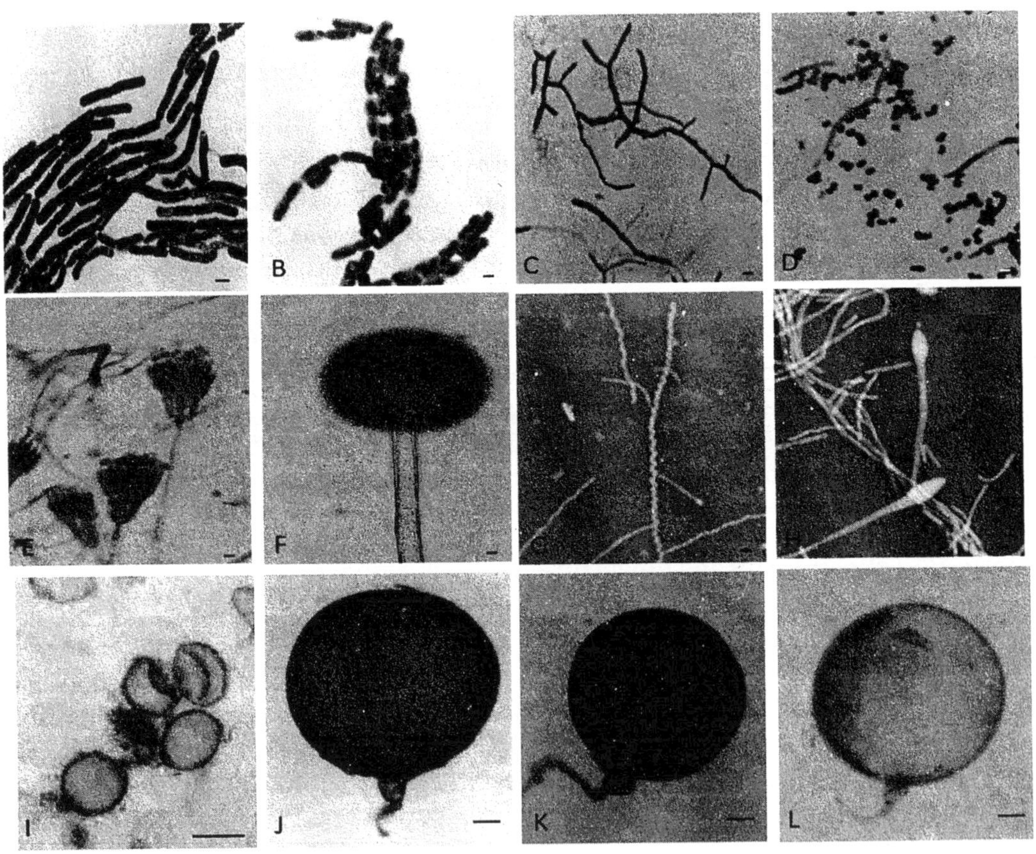

Fig. 1: Microbial diversity- *Bacillus* spp. (A&B), *Streptomyces* spp. (C&D), *Penicillium* & *Aspergillus* spp. (E&F), Thermophillic filamentous organisms (G&H), Spores of arbuscular mycorrhizae (I-L). Bar=2mm (A-H); Bar=20mm (I-L)

Thermophilic microorganisms from Hot springs

Hot springs are manifestations of geological activity that represent extreme environments that support a variety of microorganisms, filamentous organisms in particular. Occurrence of a pink filamentous community in the Octopus Spring and a black filamentous community in association with the thermal springs at Calcite in Yellow Stone National Park, USA, has been reported (Reysenbach *et al.*, 1994; 2000). Microbial analyses of soil samples collected from two hot springs (Soldhar and Ringigad) located in the Garhwal region of Uttaranchal Himalaya exhibited dominance of bacteria, in addition to filamentous forms (Fig.1.G&H). A total of 59 morphologically distinct isolates (58 aerobic and one anaerobic) were isolated and developed as pure cultures. The isolates were grouped as thermotolerants, thermophiles and hyperthermophiles. High temperature, alkaline nature and poor nutrient status of the soil prevailing at the hot spring sites seem to provide a unique environment for the development of specialized microbial communities (Kumar *et al.*, 2004).

Mycorrhizal associations in Himalayan trees

Mycorrhizae are beneficial associates of plants that influence plant growth in terms of increased nutrient and water absorption from the soil, protection from pathogens, increased tolerance to soil toxins, and root elongation (Trofymow and van den Driessche 1991; Podila and Varma, 2005). The status of ectomycorrhizae and their importance in plant growth in the context of Himalayan trees has been described by Lakhanpal and Sagar (1994). Detailed investigations conducted on the arbuscular mycorrhizal (AM) associations associated with *Taxus baccata* and five species of rhododendrons found in Uttaranchal Himalaya were recently reported from this laboratory. In the case of *Taxus baccata*, root and rhizosphere soil samples were collected from a number of sites in the Himalayan region, representing sub temperate to temperate, as well as the transition zone between the temperate and alpine conditions. The colonization by the AM fungi was characterized by the presence of arbuscules, inter/intra cellular hyphae, vesicles and extraradical chlamydospores. Five genera of the order-Glomales, namely *Acaulospora*, *Entrophospora*, *Glomus*, *Gigaspora* and *Scutellispora* were isolated from the rhizosphere soil (Chaurasia, 2005a). For investigating the AM associations in rhododendrons, root and rhizosphere soil samples of all the five species of rhododendrons (*Rhododendron anthopogon*, *R. arboreum*, *R. campanulatum*, *R. barbatum* and *R. lepidotum*) were collected from temperate, sub-alpine to alpine locations in the altitudinal range of 1500 to 4500 m amsl (Chaurasia, 2005b). In the cited study, the distribution of AM fungi along with details of externally associated functional structures (extraradical mycelium, vesicles and chlamydospores) as well as internal associations (intraradical hyphal coiling, intraradical vesicles and chlamydospores) was described. The AM colonization in root samples ranged from 28-43%, and the maximum and minimum colonization were observed in *R. arboretum* and *R. lepidotum*, respectively. Spore populations were found to belong to five genera, namely- *Acaulospora*, *Glomus*, *Gigaspora*, *Sclerocystis* and *Scutellispora* (Fig. 1I-L).

Biotechnological applications

While major efforts have focused on generating the basic information on the occurrence, isolation and characterization of various microbial communities in different niche areas spread across the Himalayan region, selected pure cultures have been investigated for possible biotechnological applications. Screening for the desirable traits and selection of efficient strains has been carried out continuously in the laboratory. For example, development of microbial inoculants for improving the plant productivity, and investigations related to the production of secondary metabolites and important enzymes have resulted in useful findings of applied value. Based on initial petridish screenings, bioassays and field studies, selected species of *Bacillus* and *Pseudomonas* have been developed in form of carrier based bioinoculants, suitable for application in the relatively colder regions (Pandey *et al.*, 1998; Bisht *et al.*, 2003; Pandey *et al.*, 2004; Trivedi, 2005). Occurrence of an antagonistic and phosphate solubilizing bacterial strain of *Pseudomonas putida* (B0) isolated from a sub-alpine Himalayan forest site has recently been reported (Pandey *et al.*, 2006). The extremophiles, thermophiles in particular, have been a subject of extensive investigation, with a view to the production of industrially useful enzymes, such as amylase, protease, cellulase, etc., (Pennisi, 1997; Srinivasan and Ingale, 1999; Budhiraja *et al.*, 2002). Kinetics of amylase activity of *Sacharomycopsis fibuligera*, isolated from the hot spring site in Garhwal Himalaya has been recently reported, from this laboratory (Kumar *et al.*, 2005).

Culture collections and the conservation of microbial diversity

The idea of establishing culture collections was conceived during the period 1880-1890, at a time when solid culture media such as those based on potato, gelatin, and agar were being devised. This led to the isolation of pure cultures of microorganisms, and at this stage, need for the "preservation" of such pure cultures for future work was also felt. This is important from the point of view of the ready availability of basic microbial resource for carrying out fundamental studies, as well as for providing the raw material for developing novel products. Culture collections have responded to this challenge by providing repositories for the safe "live" storage of germplasm, including nucleic acid sequences, and gene banks, thereby ensuring the maintenance of all species as well as the endangered ones. It must be understood that large amounts of biological materials need not be stored physically, in the future. It may be adequate only to store tissues, small samples of soil, etc., as frozen "blueprints" or micro-ecosystems (Sly *et al.*, 1990).

The cultures held in the world's microbial culture collections represent the known culturable microbial diversity. Almost one million cultures are known to be preserved and maintained in the major collections, and considerably more are likely to be held by research scientists and the industry in private or personal collections. These cultures represent a valuable genetic resource (Sly, 1994). Microorganisms often require special preservation methods in order to ensure optimal viability, storage, purity and stability of individual strains. A culture collection of microbial isolates isolated from various locations in the Indian Himalayan region has been developed in this laboratory. Following preliminary characterization, 'the microbial isolates' are regularly being accessioned in the National or International repositories. Bacterial, actinomycetes and yeast cultures are being accessioned by the Microbial Type Culture Collection and Gene Bank, Institute of Microbial Technology, Chandigarh, and fungal cultures by the Indian Type Culture Collection, Indian Agricultural Research Institute, New Delhi. Two of the plant growth promoting rhizobacteria, *Bacillus subtilis* and *Pseudomonas corrugata,* have been accessioned at the Agricultural Research Culture Collection, International Depository Authority, Illinois, USA (Accession nos. NRRL B-30408 and NRRL B-30409, respectively). With continued explorations on extreme environments, *e.g.*, for the isolation and maintenance of thermophilic microorganisms under laboratory conditions, special requirements in terms of particular conditions for incubation, modified growth media and high incubation will be required. For maintenance of thermophiles, for example, agar flakes (containing the organism) were developed by prolonged incubation at the optimum temperature in agar plates (agar concentration 2%). The cultures could be revived easily by introducing small piece of dried agar on fresh medium (Kumar *et al.*, 2004). The pure isolates of mesophiles and psychrophiles are being maintained by regular subculturing and deep-freezing in 10% glycerol at -20°C.

Conclusion

The documentation and conservation of biodiversity including microbial diversity is important as it provides the potential source of biological resources. Microbial culture collections are well recognized as germplasm banks which contribute resource pools for biotechnological research and development (Srinivasan, 1992). Through application of molecular methods, such as direct extraction of nucleic acids from the environment, application of PCR (Polymerase chain reaction) technologies and sequence analysis, the assessment of the biodiversity of prokaryotes including culturable and non-culturable as well, is being taken up. The identity of a microbial species can be assessed by analysis of its genetic material, preferably by 16S ribosomal DNA (Stackebrandt and Liesack, 1993). Initiatives on exploration, documentation and preservation of microbial diversity of Himalayan region have been taken up by this laboratory with special reference to their biotechnological applications. While initially the characterization of microorganisms was being conducted on the basis of their morphological, biochemical and physiological parameters, the molecular characters of the microorganisms has also been included.

References

Bisht, D., Pandey, A. and Palni, L.M.S. 2003. Influence of microbial inoculations on *Cedrus deodara* in relation to survival, growth promotion and nutrient uptake of seedlings and general soil microflora. *Journal of Sustainable Forestry,* **17**(3): 37-54.

Budhiraja, R., Basu, A. and Jain, R.K. 2002. Microbial Diversity: Significance, conservation and application. *Science Letters.* **25**(7&8):189-201.

Chaurasia, B., Pandey, A. and Palni, L.M.S. 2005a. Occurrence of Arbuscular Mycorrhizae in the rhizosphere of Himalayan Yew (*Taxus baccata* L. subsp. *wallichiana* (Zucc.) Pilger)-A Case Study. In: Basic Research and applications of Mycorrhizae Eds. GK Podila and AK Varma, 26-35, IK International Pvt. Ltd., New Delhi.

Chaurasia, B., Pandey, A. and Palni, L.M.S. 2005b. Distribution, Colonization and Diversity of Arbuscular Mycorrhizal fungi in Rhododendrons of Central Himalayan region of India *Forest Ecology & Management.* **207**(3): 315-324.

Kumar, B., Trivedi, P., Mishra, A.K., Pandey, A. and Palni, L.M.S. 2004. Microbial Diversity of soil from two Hot Springs in Garhwal Himalaya. *Microbiological Research,* **159**(2): 141-146.

Kumar, B., Pandey, A. and Palni, L.M.S. 2005. Extracellular amylase activity of *Saccharomycopsis fibuligera*, a mycelial yeast isolated from a hot spring site in Garhwal Himalayas. *Indian Journal of Microbiology,* **45**(3): 211-215.

Lakhanpal, T.N. and Sagar, A. 1994. The Ectomycorrhiza-current status. *Indian Journal of Microbiology,* **34**: 1-26.

Pandey, A. and Palni, L.M.S. 1998a. Microbes in Himalayan Soils: Biodiversity and Potential Applications. *Journal of Scientific and Industrial Research,* **57**: 668-673.

Pandey, A. and Palni, L.M.S.1998b. Isolation of *Pseudomonas corrugata* from Sikkim Himalaya. *World Journal of Microbiology and Biotechnology,* **14**: 411-413.

Pandey, A., Sharma, E. and Palni, L.M.S. 1998. Influence of bacterial inoculation on maize in upland farming systems of the Sikkim Himalaya. *Soil Biology and Biochemistry,* **30**: 379-384.

Pandey, A., Trivedi, P., Kumar, B., Chaurasia, B., Singh, S. and Palni, L.M.S. 2004. Development of microbial inoculants for enhancing plant performance in mountains. In: Biotechnological Approaches for Sustainable Development. Eds. M.S. Reddy and S. Khanna, pp. 7-20, Allied Publishers Pvt. Limited, New Delhi.

Pandey, A., Trivedi, P., Kumar, B. and Palni, L.M.S. 2006. Characterization of a phosphate solubilizing and antagonistic strain of *Pseudomonas putida* (B0) isolated from a Sub-Alpine location in the Indian Central Himalaya. *Current Microbiology* (In Press).

Pennisi, E. 1997. In industry, extremophiles begin to make their mark. *Science,* **276**: 705-706.

Podila, G.K. and Varma, A. (Eds) 2005. Basic Research and Applications of Mycorrhizae, IK International. Pvt. Ltd., New Delhi.

Reysenbach, A.L., Wickham, G.S. and Pace, N.R. 1994. Phylogenetic analysis of the hyperthermophilic pink filament community in Octopus spring, Yellow Stone National Park. *Applied and Environmental Microbiology,* **60**(6): 2113-2119.

Reysenbach, A.L., Ehringer, M. and Hershberger, K. 2000. Microbial diversity at 83 °C in calcite springs, Yellow Stone National Park: Another environment where the Aquificales and "Korachaeota" coexist. *Extremophiles,* **4**: 61-67.

Satyanarayana, T., Raghukumar, C. and Shivaji, S. 2005. Extremophilic microbes: diversity and perspectives. *Current Science,* **89**(1): 78-90

Seackbach, J. 2000. Journey to Diverse microbial worlds, Vol. 2, Kluwer Academic Publishers, Dordrecht.

Sly, L.I., Lijima, T. and Kirsop, B. (Eds). 1990. 100 Years of Culture Collections, Proceedings of the Kral Symposium to celebrate the centenary of the first recorded service culture collection, International House, Osaka, Japan.

Sly. L.I. 1994. Culture Collections World-wide. The Biodiversity of microorganisms and the role of Microbial Resource Centres. Eds. B Kirsop and DL Hawksworth, pp. 29-35, WFCC Biodiversity Committee, UK.

Srinivasan, M.C. 1992. Reflections of Endangered Culture Collections and their rescue, with special reference to developing countries. In: Endangered Culture collections, Proceedings of the First and Second International Symposia organized by the World Federation for Culture Collections.

Srinivasan, M.C. and Ingale, S. 1999. Production and application of enzymes stable to and active under extreme environments: an overview, PINSA, **b65**(3&4): 143-162.

Stackebrandt, E. and Liesack, W. 1993. Nucleic acids and classification, Handbook of New Bacterial Systematics, 151-194. Academic Press, London.

Stolp, H. 1988. Microbial Ecology: Organisms, Habitats, Activities. Cambridge University Press, Cambridge.

Trivedi, P., Pandey, A. and Palni, L.M.S. 2005. Carrier based formulations of plant growth promoting bacteria suitable for use in the colder regions. *World Journal of Microbiology and Biotechnology,* **26**(6&7): 941-945

Trofymow, J.A. and van den Driessche, R. 1991. Mycorrhizas in mineral nutrition of conifer seedlings (ed.) van den Driessche R, CRC Victoria, BC Canada, pp. 183-225.

Soil Microbiology and Soil Biotechnology : Vol. 1
Editor : S. Kannaiyan
Associated Publishing Co., New Delhi, 2010, pp. 136-157

10

Culturable and Non-Culturable Approaches to Study Microbial Diversity in Soil Ecosystem

*Shilpi Mittal and Jugsharan S. Virdi**

Introduction

The most biological diversity on this planet is probably harbored by the soil. The soil microorganisms (bacteria, fungi, algae, worms and nematodes) play pivotal roles in various biogeochemical cycles (BGC) and are responsible for the cycling of organic compounds, and in agriculture production. The soil microorganisms also influence the above-ground ecosystems by contributing to plant nutrition, plant health, soil structure and soil fertility to maintain natural soil ecosystem. Microbial diversity is an index of the community that takes into account both species richness and the relative abundance of the species. Diversity of an ecosystem is composed of two elements: *richness* and *evenness*, so that the highest diversity occurs in communities with many different species present (richness) in relatively equal abundance (evenness). The richness and evenness of bacterial communities reflect selective pressures that shape diversity within communities.

Soil, the heart of agroecosystem, sustains an immense diversity of microbes. Microbial diversity in soil ecosystem exceeds by several order compared to that of eukaryotic organisms. It has been indicated that a gram of soil could harbor upto 10 million microorganisms constituting possibly of thousands of different species. Among these, bacteria are most numerous microbial groups in soil, but because of their small size (1-10 mm) bacteria account for less than half of the total biomass in agricultural soil. Torsvik *et al.* (1996) have calculated presence of about 6,000 different bacterial genomes in a gram of soil, taking into account the genome size of *Escherichia coli* as the unit.

Understanding diversity and function of the microbiological component of soil poses great challenges that are being overcome by the application of molecular biological approaches. A range of novel methods, most of which are based on rRNA and rDNA analyses, have uncovered part of the soil microbial diversity. The next step in the era of microbial ecology is to draw genomic, evolutionary and functional information from bacterial artificial chromosome

Department of Microbiology, University of Delhi, South Campus, Benito Juarez Road, New Delhi - 110 021, Email: virdi_dusc@rediffmail.com libraries of the soil community genomes (the metagenome). Sophisticated analyses that apply molecular phylogenetics, DNA microarrays, functional genomics and *in situ* activity measurements have provided huge amounts of new data, potentially increasing our understanding of the structure and function of soil microbial ecosystems, and the interactions that occur within them.

Culture dependent assessment of bacterial diversity

Direct plate count and the Great Plate Count Anomaly

Viable plate counts that have been used for decades to assess the culturable microflora in environmental samples restrict growth of most of the microorganisms. In fact, microorganisms isolated using standard cultivation methods are rarely numerically dominant in the communities from which they were obtained: instead, they are isolated by virtue of their ability to grow rapidly into colonies on high-nutrient artificial growth media, typically under aerobic conditions, and at moderate temperatures. Easily isolated organisms are the 'weeds' of the microbial world and are estimated to constitute less than 1% of all microbial species. Staley and Konopka (1985) coined the phrase, "great plate count anomaly" to describe this phenomenon. It has been frequently reported that microscopic counts exceed viable-cell counts by several orders of magnitude. Direct counting by fluorescence microscopy is reported to give 100-1000 times more than the number obtained by plate counting since this technique does not differentiate between dead and living cells.

Phenotypic assessment

The classical phenotypic characterization of bacteria comprises morphological, physiological and biochemical features. The morphotypes of bacterium include both cellular (shape, endospore, flagella, inclusion bodies, gram staining) and colonial (color, dimensions, forms) characteristics. The physiological and biochemical features include data on growth at different temperature, pH, salt concentration or atmospheric conditions, and growth in the presence of various substances such as antimicrobial compounds. The fluctuations and deviations in phenotypic characters described normally are due to visual interpretation of specimens and samples by different workers.

A number of methods, which have greater discriminative power and are less time consuming, have now been developed for characterization of microbial communities. In the category of phenotypic characterization, the new technique based on metabolic fingerprinting of bacterial isolate is typified by BIOLOG™. This technique utilizes 95 different substrates (carbohydrates, amino acids and organic compounds) as sole carbon sources. Variations of the commercially available BIOLOG™ microbial identification system are now widely available to assess functional diversity of microorganisms from environmental samples.

Cellular fatty acid profile is one of the most widely accepted approaches used to study the microbial population diversity. The cellular fatty acids are used as the chemotaxonomic marker. This is based on the lipid content present in bacterial cells. Polar lipids are major constituents of the lipid bilayer of cell membrane and fatty acids are the major constituents of lipid and lipopolysaccharide present in bacterial cells. The fatty acids are analyzed as methyl esters and therefore the technique has been termed as "fatty acid methyl ester" (FAME) analysis. For the extraction of fatty acid methyl esters, each sample is first saponified followed by methylation and then extracted into hexane and methyl-tert butyl ether (MTBE). The fractions obtained are washed with alkali before concentration overnight by volatilization of the hexane and MTBE. The extracted samples are analyzed on a gas chromatograph. The instrument is programmed and controlled with MIDI Sherlock Microbial Identification System software (Version 3.0) (MIDI, 1999) using MIDI microbial calibration standards (#1 quantitative saturated nC:9 to nC20:0 plus 2 and 3 Hydroxy). Fatty acid methyl esters are quantified based on these calibration standards. Fatty acids are denoted using MIDI nomenclature and were designated by the number of carbon atoms present, a colon followed by the number of double bonds, and then the position of the first double bond from the omega (w) end of the molecule. More than 300 different chemical structures of fatty acids have been identified. The variability in chain length, double bond position and substituent groups from organisms to organisms is helpful in categorizing microorganisms and assessing their diversity.

Fourier-Transform Infrared (FT-IR) spectroscopy

Although not a very new technique, FT-IR has been developed in the last few years as a new tool for rapid identification and diversity of clinically relevant microorganisms. Bacterial fingerprints based on FT-IR are highly reproducible and typical of different species, and can be used for probing the identity of microorganisms. The simplicity and versatility of FT-IR spectroscopy makes its application in rapid identification, classification, differentiation, and large scale screening of isolates at subspecies level a reality (Cho-Smit *et al.,* 2001). Since this technique utilizes vibrational characters of all cell constituents, *i.e.*, nucleic acid (DNA/RNA), proteins, membranes and cell wall components, it displays phenetic as well as genetic fingerprint of a microbe. Use of FT-IR avoids tedious procedure of sample preparation since in this technique either a film of cell suspension on Zn-Se optical slide is used or the cell mass (~10-60 μg) is placed on an IR-sample holder for spectral analysis; the processing time is therefore considerably reduced and within a day >100 samples can be analyzed. FT-IR is considered as sensitive and able to discriminate microorganisms up to higher level (sub species level), which can be influenced by the physicochemical parameter (pH, CO_2 concentration, dissolved O_2 etc.) effecting bacterial cell growth.

Molecular assessment

Apart from phenotypic characterization, a number of molecular methods with high resolving power to characterize microbial community are now available. This has been achieved through advancements in genomic techniques. Molecular characterization helps explore the microbial diversity with greater discrimination and provides a benchmark to develop new technologies to explore hidden microflora of a system. The information generated from nucleic acid content of cell is utilized for genotyping. The molecular methods are based mainly on isolation of nucleic acid (DNA) from organisms of interest, which are then analyzed using either PCR-independent methods (genomic DNA-DNA hybridization, G+C content analysis etc.) or by PCR-dependent methods such as DNA fingerprinting by amplified rDNA restriction analysis (ARDRA) and repetitive-PCR (rep-PCR), terminal restriction fragment length polymorphisms (T-RFLP) and 16S rDNA sequencing.

The generally applied measure of sequence divergence employs the use of DNA-DNA hybridization to evaluate total genomic similarity. This technique is one of the most important tools in microbial systematics and has been applied to evaluate environmental diversity. The percent DNA binding or DNA-DNA hybridization value or the relative binding ratio is an indirect parameter of the sequence similarity between the two entire genomes. In addition to DNA-DNA hybridization, low-resolution methods include analysis of base distribution (mole percent guanine + cytosine) of DNA and determination of rate at which denatured single stranded DNA reassociates when temperature is lowered to approximately 25°C below the DNA melting point (Torsvik *et al.,* 1996).

PCR-dependent molecular approaches

Most of the PCR dependent diversity measures are based on the use of 16S ribosomal RNA (rRNA) molecule.

- **Why rRNA genes?** rRNA operon is transcribed into one pre-RNA transcript that contains three *i.e.*, 16S, 23S, and 5S rRNA (Fig. 1a). Several features of rRNA operon make it a "molecular chronometer" in assessing population diversity. The sequences that code for rRNA genes are among the most highly conserved (Woese, 1987). Ribosomal RNA is a structural domain in which sequence variation increases with increase in phylogenetic distance. This attribute favours its use as a 'molecular chronometer'. Regional differences in sequence conservation help in nucleic acid probe and primer design for various specificities, based on which group and species-specific oligonucleotide probes/primers have been used for direct assessment of environmental diversity.
- **Why only 16S rRNA gene?** The 16S rRNA region comprises of approximately 1,500 nucleotides and provides considerable information for phylogenetic analysis whereas 23S rRNA gene region (~ 3,000 nucleotides) generates twice as much information as 16S rRNA. While 16S rRNA molecule is not only sufficiently large to generate enough information, it is much easier to sequence. Therefore, it has helped establish a powerful tool and has strong support in its universal acceptance. 5S rRNA gene region (~120 nucleotides) is very small and contains little information and thus has been less useful in the study of distant phylogenetic relationships.

The designing of primers for specific applications, and data on rDNA sequences can be obtained through Ribosomal Database Project (RDP-II: http://www.cme.msu.edu/RDP1). To design universal primers, sequence of region 1 and 2 and for specific primers 3 and 4 (Fig. 1b) can be used (Louws *et al.,* 1999).

Amplified rDNA restriction analysis (ARDRA)

This is an easy and rapid method to characterize bacterial diversity (richness and evenness) based on 16S rDNA. The strategy adapted in this tool requires the following sequential steps: (i) grouping of isolates recovered into clusters corresponding to bacterial species on the basis of restriction patterns obtained after PCR amplification. DNA digestion is performed using tetrameric restriction enzymes (*Alu*I, *Hae*III *etc*), which generate species-specific restriction pattern (ii) strains are assigned to species by comparing the restriction pattern with those of reference strains, if a particular member of a community is to be identified. ARDRA is one of the most widely accepted molecular tools used by microbial ecologists to study culturable diversity of an ecosystem.

Internally Transcribed Spacer (ITS)-PCR

The rRNA operon is transcribed into pre-rRNA transcript that contains the following components in the order 5' to 3': 16S-spacer-tRNA-spacer-23S-spacer-5S (Fig. 1a). Sequence information has also become available on the 16S-23S spacer region and suggests that considerable variation can occur between species in both the length and sequence of this region because of the presence of several tRNA genes and non-coding regions that appear to be under less selection pressure. Which makes ITS region more variable than 16S and 23S rRNA genes. The presence of multiple copies (alleles) of rRNA operon per genome supports the utilization of these sequence variation as a powerful tool to study species diversity. Conserved regions flanking the 16S and 23S rRNA genes serve as primer recognition sites for PCR amplification (region 5 and 6 of Fig. 1b) of the ITS region followed by restriction digestion with tetrameric endonucleases. To increase specificity of ITS amplification, two step PCR was performed, first amplification is carried out by amplification employing primers designed to anneal full 16S rRNA gene followed by amplification using primes that anneal internal to the ITS region.

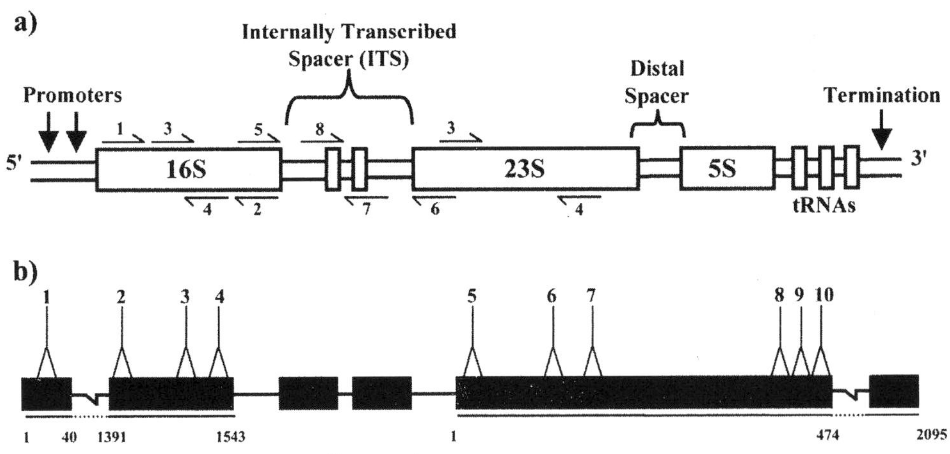

Fig. 1a: *Organization of ribosomal RNA operon in bacteria:* The arrows indicate the regions the sequence of which are used to design universal/specific primer set

Fig. 1b: *Positions of conserved regions (1-10) within the rRNA operon:* The dark boxed areas represent the various genes of the bacterial rRNA operon. The dark line represents the spacer regions that separate the various rRNA genes (Louws *et al.*, 1999)

Genus/Group specific diversity measure

To facilitate genus and species level identification of a broad range of bacteria coupled to presumptive identification, unified set of primers have been developed which correspond to either whole 16S rRNA gene, or spacer region between 16S and 23S genes in a prokaryotic rRNA genetic locus. The utility of rRNA gene sequence as a taxonomic tool has been amply demonstrated in bacteria; 16S rRNA gene sequence analysis has, in fact, completely redefined the phylogenetic relationships. In addition to highly conserved areas that have been used to study the relationship among distant taxa, the 16S sequences containing more variable regions have been useful in differentiation of genera and species. Identification of a pure culture can be accomplished by using specific primers (*i.e.,* universal eubacterial, genus and/or species specific) to amplify a specific gene that provides a unique genomic fingerprint.

The rRNA gene approach to phylogenetic identification and *in situ* detection of microorganisms has been most widely used by molecular microbial ecologists. Oligonucleotide probes are more suitable for routine screening such as signature molecules. Such probes have been designed for the differentiation of highest taxonomic level, the domain Archaea, Bacteria and Eukarya; for intermediate levels, *e.g.*, the gram-negative sulphate reducing bacteria, the α, β and γ subclass of proteobacteria, and the *Flavobacterium-Cytophaga-Bacteroides* clusters of the CFB phylum which include lower taxa (genus, species and subspecies specific probe) also. In addition to the use of specific probes for taxonomic studies, use of specific primer sets that target various unique genes is in use for the last more than ten years. Genus specific rDNA sequences among phytobacteria are well documented and primers have been developed for pseudomonads, xanthomonads and several other genera. The rRNA genes can also function as targets for highly selective detection of target DNA, but the level of discriminatory power lies at the species or genus levels. The universal nature of many such PCR primers makes them highly adaptable for devising other PCR-based strategies.

Repetitive (rep)-PCR

In contrast to the analysis of small part of genome such as 16S rDNA region, whole genomic DNA is also used to study diversity based on repetitive extragenic palindromic sequences based PCR fingerprinting methods (Fig. 2) (Versolovic *et al.,* 1994). Both, prokaryotic and eukaryotic genomes contain dispersed repetitive sequences separating larger single copy DNA sequences. Various classes of repeated DNA sequences are present in diverse prokaryotic genomes. The interspersed repetitive DNA sequences range between 15 to several hundred base pairs in length. These elements are non-coding but are present in high copy number relative to the longer repeated elements, which contain coding sequences. The interspersed repetitive sequences described in bacteria are Box element (154 bp), repetitive extragenic palindromic sequences (REP) sequences (386 bp) and enterobacterial repetitive intergenic consensus (ERIC) sequences (124 bp). Genomic fingerprints with the sequence-based primers distinguish unrelated bacterial populations because individual bacterial strains vary with respect to the distance between the two repetitive sequences. REP-PCR utilizes outwardly facing oligonucleotide primers, complementary to the interspersed sequences. These primers amplified DNA fragments of different size consisting of the region lying between the REP-elements (Fig. 2).

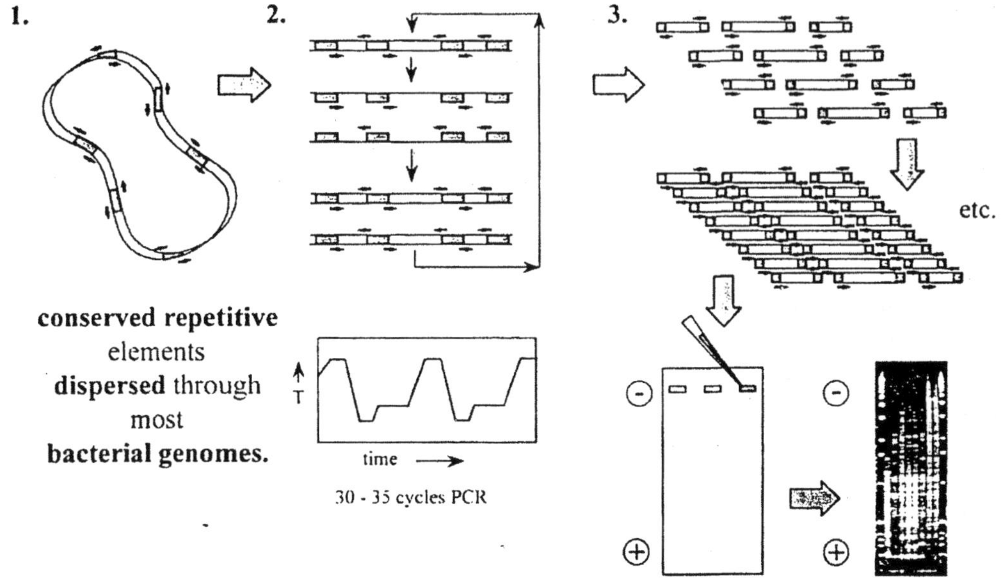

Fig. 2: Outline of rep-PCR genomic fingerprinting technique (Louws *et al.*, 1999)

Different sized amplicons can be visualized by electrophoresis for typing of bacterial strains. rep-PCR fingerprinting has been used to correlate the phenotypic grouping of organisms with genetic relatedness, and to group large gene pool of organisms into different clusters. With few exceptions, comparison of rep-PCR fingerprinting has been reported to correlate well with that resulting from phenotypic analysis.

Sequencing

The above story is however incomplete without sequencing because this tool provides much greater discrimination than other methods and characterizes particular members of a community better. Results of comparative 16S ribosomal RNA sequence analysis have made study of diversity much simpler since the first reports of Pace *et al.* (1986) and Woese and Fox (1977). In general, sequencing of the 16S ribosomal DNA products is done on the Genetic Analyzer. The instrument is based on the principle of capillary electrophoresis (CE) that uses a laser beam to excite fluorescently labeled DNA fragments as they pass the laser window. The fragments are rapidly separated (>450 bases at 98% accuracy). The sequences are aligned and direct identification is made using the 16S ribosomal RNA database, which currently contains sequences from approximately 8,000 organisms.

The phylogenetic approach to systematics and assessment of culturable microbial diversity upto taxonomic level using nucleic acid hybridization and 16S rRNA gene sequences analysis has been of much utility in the phylogenetic reformation of the classification of prokaryotic organisms. Much knowledge is now available that permits use of DNA sequences to identify, and understand the distribution and role of microbial population in the environment (Ward *et al.*, 1998). Several workers have used 16S rDNA cloning and sequencing of DNA is currently being used to study the community composition of an ecosystem (Schwieger and Tebbe, 2000).

Culture independent assessment of microbial diversity

In spite of the many methods being regularly used for assessing culturable diversity as described above, available evidence shows that microorganisms in nature are more diverse phylogenetically than has been accounted for by sequence analysis of cultured strains. It has been estimated that 99% of the bacterial species found in the soil cannot be cultured in the laboratory. The indication of vast diversity of uncultured life in soil has stimulated development of culture independent methods for study of microbial diversity, to avoid the bias caused by cultivation under laboratory conditions. In recent years, several cultivation independent community level targeting techniques have been developed that can be used as an adjunct to traditional methods. These employ a combination of nucleic acid characterization along with other macromolecules such as the use of phopholipid fatty acid analysis and fatty acid methyl ester analysis. For characterization of complex and immense diversity of microbial populations present in soil and rhizosphere, the following approaches are generally being applied.

Functional diversity

Of late, species units based on functioning or process attributes have been used as a diversity measures. Substrate utilization through 95 well BIOLOG microplate has been used to differentiate species unit based on the ability of bacteria to metabolize various carbon substrates (Kennedy, 1994). The use of BIOLOG microplate for the estimation of functional diversity of uncultured microorganisms and/or natural *in situ* soil sample are termed as community level physiological profile (CLPP), which provides information about the combined catabolic potential or potential activity of a system. Potential activity is the capability of the community to adapt its metabolism (catabolism) and/or relative composition and size of constituent populations to varying abiotic conditions.

Structural diversity

Analysis of structural component of an ecosystem is used to distinguish its species and diversity spectrum. The structural diversity measurement can be arrived through the use of lipids in the cell, which is characteristic of a species but is influenced by environmental perturbations. Fatty acids can be used to fingerprint bacterial species and soil through fatty acid methyl ester (FAME) and phospholipid fatty acid (PLFA) analysis. Whole cell FAME analysis detects lipids from both live and dead microorganisms, other macroorganisms and humic materials. Nonetheless, the living microbiological component of soil community can be estimated by PLFA analysis as these structural constituents are found only in the membranes of living cells. These diversity measures, however, cannot generate taxonomic information on individual species or groups present in the soil. These techniques have been used successfully to assess the effect of soil type on microbial community structure, to characterize microbial population dynamics under different cropping system, effect of soil depth, soil management practice (tillage system) under continuous cotton cropping system, and plant type and soil type on soil microbial community structure. Use of FAME profiles of soils subjected to different agricultural management practices revealed different FAME profiles indicating management-induced changes in the composition of microbial communities.

Genetic diversity

Genetic diversity of an agroecosystem can be assessed through nucleic acid analysis of direct soil sample. Genetic analysis, most of which is based on rRNA gene, has identified several thousand independent genomes in a gram of soil. The recovery of genetic information from soil samples provide not only the relative numbers of diverse soil organisms, but have also opened up new vistas of life, such as found in kingdom Archaea. Genetic diversity studies can out compete most of the drawbacks of other techniques since genetic tools provide one of the most powerful adjuncts to all culture dependent tools. Various functional genes such as catabolic genes are often localized on mobile genetic elements susceptible to high mutation rates and horizontal transfer. Polymerase chain reaction-based methods have been successfully applied for detecting catabolic genes in the environment. Such methods, however, have focused mainly on the characterization of the presence, or abundance of a family of catabolic genes without assessing the finer levels of sequence diversity. Most of the genetic diversity measures utilize polymerase chain reaction based 16S rDNA profile that provide higher resolution and generates information about changes in the community structure. These techniques, which include denaturing gradient gel electrophoresis (DGGE), single strand conformation polymorphism (SSCP), terminal restriction fragment length polymorphisms (T-RFLP), ribosomal intergenic spacer analysis (RISA), can provide detailed information about community structure of an ecosystem in terms of richness, evenness and composition.

Terminal restriction fragment length polymorphism (T-RFLP)

T-RFLP of PCR amplified DNA is an advanced fingerprinting technique based on RFLP. The general steps includes PCR amplification of a conserved target sequence [most commonly a region of the small subunit RNA (ssuRNA) gene] followed by restriction enzyme digestion with single endonuclease and gel fractionation of resulting fragments by automated DNA sequencer or a capillary electrophoresis system. In T-RFLP, one of the two 16S rDNA primers is fluorescently labeled at 5' end, which results in PCR product tagged with a fluorescent dye at only one terminal. Gene encoding the rRNA sequence is the most commonly used marker in microbial community analysis by T-RFLP. A growing number of studies have demonstrated that T-RFLP is a sensitive, reproducible, and robust methodology for comparative analysis of an environmental sample. Relative diversity of complex microbial communities in four soil environments was investigated employing T-RFLP profile of 16S rRNA genes and consistency of results was checked by comparison with other methods (Dunbar *et al.,* 2000). The results indicated that T-RFLP method was able to differentiate the four communities, which were also consistent with the previous results of 16S rDNA clone library.

SSCP (Single-strand conformation polymorphism)

SSCP separates ssDNA fragments differing in strand length and conformation (composition) by gel electrophoresis. SSCP-PCR was initially used to detect minor sequence variations caused by point mutations in human alleles and has recently been used to analyze microbial community structure (Zumstein *et al.*, 2000). Prior to SSCP analysis, the PCR amplified rDNA is denatured at high temperature in presence of denaturants followed by rapid cooling on ice to "trap" different structural forms in DNA. The folding of DNA into different stable structure is highly related to its sequence (*i.e.*, nucleotide sequence and composition), which is dependent upon the sequences of different rDNA fragments of a community that can be resolved on the basis of varying migration rates during electrophoresis. Mainly three different conformational types are observed, *viz.*, self-folded ssDNA, heteroduplex double-stranded DNA (dsDNA), and homoduplex double stranded DNA (dsDNA). Differentiation between dsDNA and ssDNA is based upon relative migration distance and by variation in silver stain color (Liu and Stahl, 2000). Effect of inoculation on indigenous microflora has been investigated using SSCP, which revealed that plant species and the inoculant affect the structural diversity in the rhizosphere (Schwieger and Tebbe, 2000). In another study, bacterial community in rhizosphere of transgenic maize was compared to its isogenic, non-transgenic cultivars and it was found that neither the genetic modification nor the use of herbicide affected the SSCP profiles whereas the rhizosphere community of sugarbeet, grown in same field as a control crop, showed clearly different SSCP profile of bacterial community (Schmalenberger and Tebbe, 2002). The technique has also been used to study functional diversity of biodegrading community in contaminated soil.

DGGE and TGGE

Denaturing gradient gel electrophoresis (DGGE) and temperature gradient gel electrophoresis (TGGE) were first developed to detect single base change in genes for the diagnosis of human genetic diseases and in genetic linkage studies. These techniques were introduced into microbial ecology by Muyzer *et al.* (1993). DGGE and TGGE are based upon analytical separation of DNA fragments of identical or near-identical lengths but with varying sequence compositions. Separation is based on decreased electrophoretic mobility of a partially melted double-stranded DNA molecule in polyacrylamide gel containing a linearly increasing gradient of DNA denaturants (a mixture of urea and formamide) or a linear temperature gradient. A mixture of PCR products obtained with genomic DNA extracted from a complex assemblage of microorganisms and primers specific for a molecular marker, such as the 16S rRNA gene, is separated in a polyacrylamide gel containing a linear gradient of DNA denaturants. Sequence variation among the different DNA molecules influences the melting behavior, and therefore molecules with different sequences will stop migrating at different positions in the gel. Changes in fragment mobility are associated with partial melting of the dsDNA in a discrete region, termed as 'melting domain'. For example, in DGGE analysis under a constant temperature, the Tm of each domain varies according to the concentration of denaturant, and therefore different positions in the gel. When DNA enters a region of the gel containing sufficient denaturant for partial melting of domain, a transition of the helical to partially melted molecule occurs and migration is severely retarded at that positions in the gel during migration. Sequence variation within such domain causes change in melting temperature, and molecules with different sequences stop migrating at different positions in the denaturing gel. A sequence of guanines (G) and cytosines (C) is added to the 5' end of one of the PCR primers, co-amplified during PCR and is thus incorporated into the amplified DNA fragment. The GC rich sequence acts as a high melting domain preventing the DNA strand from complete denaturation into single strand. The length of 'GC clamp' can very between 30 and 50 nucleotides.

The information about the species composition can also be obtained from very complex microbial communities by excision, cloning and sequencing of desired band from the DGGE and TGGE gels.

Metagenomics in diversity analysis: community genome analysis

Metagenomics, the genomic analysis of a population of microorganisms, has emerged as a powerful molecular tool to assess the microbial diversity. Metagenomic analysis involves isolating DNA from an environmental sample, cloning the DNA into a suitable vector, transforming the clones into a host bacterium, and screening the resulting transformants for the study of microbial community of a particular property. The clones can also be screened for phylogenetic markers or "anchors," such as 16S rRNA and *recA* to study the microbial evolution. Direct isolation of genomic DNA from an environment circumvents culturing the organisms under study, and cloning of it into a cultured organism captures it for study and preservation.

The word metagenomics was coined to describe the functional and sequence-based analysis of the collective microbial genomes contained in an environmental sample (Fig. 3). The study of community using single genome provided insight of only individual cell, whereas metagenomic approach can unlock the secrets of community life on a broader scale. The concept of cloning DNA directly from an environment was initially suggested by Pace (1985) and first implemented by Schmidt *et al.* (1991), who constructed a ë phage library from a seawater sample and screened it for 16S rRNA genes. Many environments are the focus of metagenomics that include soil as one of the important ecosystems of study. However, development of metagenomic analyses of soil was slower than with seawater because of the technical challenges of cloning DNA from the complex matrix of soil, which contains many compounds that bind to DNA or inhibit the enzymatic reactions required for cloning. Significant progress has been made, producing libraries that have substantially advanced understanding of the functions in the soil community. The metagenomic approach permits analysis of novel metabolic pathways, gene regulatory elements, genes of unknown functions, and genes for pathogenesis, virulence and drug resistance (MacNeil *et al.,* 2001). Availability of complete microbial genome sequences has made it possible to identify and characterize all genes present in a species. This information indirectly provides deeper insight into evolution of genes and species. Comparative genomic analysis using new techniques such as microarray technology have therefore become very useful in functional diversity studies that lead to tracking of highly expressed genes and genes critical in biogeochemical pathways.

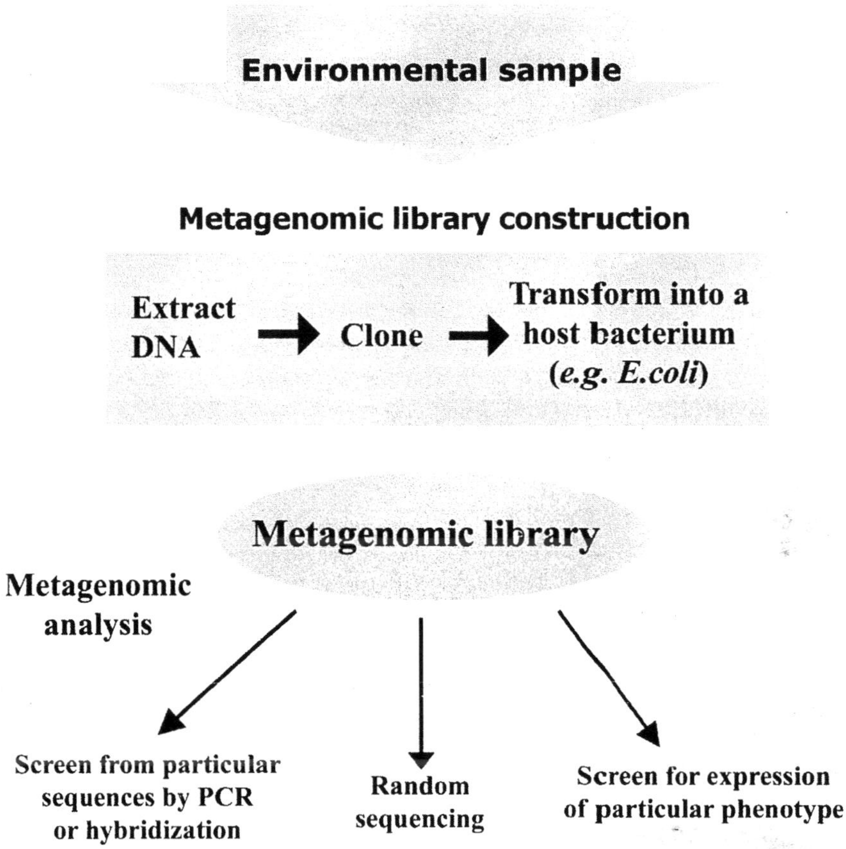

Fig. 3: Metagenomic involves construction of a DNA library from an environment's microbial population and then analyzing the functions and sequences in the library (Riesenfeld *et al.,* 2004)

DNA microarray

DNA microarray, in brief, is a miniaturized array of complementary DNA (cDNA) probes [~500 to 5,000 nucleotide (nt) in length] or oligonucleotides (15 to 70 nt) attached directly to a solid support, that permits the simultaneous hybridization of a large set of probes complementary to their corresponding DNA/RNA targets in a sample, thus allowing rapid and high throughout sequence analysis. Initially, samples containing various fluorescent-labeled target molecule are introduced into an assay containing immobilized probes of known sequences. Different targets are subsequently hybridized to the respective probes with complementary sequences. After the hybridization process, washing is carried out to remove any unattached and/or non-specific targets. Probes with attached targets emit fluorescent signals, indicating the presence of certain microbes or expressed genes in the samples. Based on the intensity of fluorescence emitted, quantification of the microbial population is also possible when the amount of a probe used exceeds that of the introduced targets, with the detected signal intensity being proportional to the amount of targets hybridized to the probes.

The quantitative description of microbial communities in terms of gene expression of particular function is currently considered as one of the most promising areas of research in microbial ecology. This has been possible through development of DNA microarray technology and its application in the study of microbial community structure in natural ecosystems (Tiedje *et al.*, 2001). Through this tool, characterization of composition and function of microbial communities is feasible since single assay can contain upto 500-1000 different spots that permit very broad hybridization with a wide identification capacity. In the different kinds of hybridization formats used in microbial community analysis, ribosomal RNA (rRNA) is the target molecule of choice. For example, comprehensive phylochips (DNA microarray consisting of rRNA-targeted oligonucleotide probes) were developed for sulfate-reducing prokaryotes (SRP-Phylochip). Bacteria of the provisional beta-proteobacterial order, "Rhodocyclales" (RHC-Phylochip) were subsequently evaluated by hybridization of fluorescently labeled 16S rRNA gene PCR amplification product with that of the selectable reference organisms. Peplies *et al.* (2003) made a successful attempt towards utilization and optimization of DNA microarray format for specific detection of bacteria based on 16S rRNA genes with a model system composed of six environmental strains and 20 oligonucleotide probes. DNA microarray technology has been used in parallel with DGGE. Consistent results were obtained through both the techniques when sulfate-reducing bacteria were characterized in toluene and ethylbenzene-degrading microbial consortia. These workers also analyzed oil contaminated sediment from the coast of Kuwait to assess the utility of DNA microarray for analysis of environmental samples. This advanced technology has also been used to monitor the 2, 4-D degrading gene expression in paper and pulp wastewater and SSU rRNA based detection of nitrifying bacteria in environmental samples. DNA microarray has also been applied for the detection of functional genes involved in the nitrogen cycle in the environment. Differences in the apparent distribution of nitrate reductase (*nirS* and *nirK*) genes, ammonia mono-oxygenase (*amoA*) gene and methane mono-oxygenase (*pmoA*) gene can be made between marine sediment samples and soil samples.

Direct rRNA based analysis of community analysis

In conjunction with DNA microarray, oligonucleotide microarray has been used to profile directly the extracted rRNA from environmental microbial population without PCR amplification. The ability to determine target rRNA sequences without amplification is likely to further improve the capacity to detect relative number of sequences representing natural microbial populations. Metal and sulfate reducing bacteria of the genera *Geobacter* and *Desulfovibrio* was detected employing an oligonucleotide array consisting of universal and species-specific 16S rRNA probes. This suggests that such methods are equally applicable to direct detection and characterization of 16S rRNA in microbial species and environmental samples.

While studies of RNA mainly provide details of community structure, specific analysis of mRNA expression provides additional information on the activities of a particular population in a sample. However, only a small number of workers have reported successful analysis of mRNA isolated from soil or sediment (Hurt *et al.*, 2001). Analysis of *nifH* mRNA expression in soil to detect nitrogen-fixing activity of diazotrophic population was reported the estimates of *nifH* gene expression level and its regulation directly in the soil environment.

Fluorescent *in situ* hybridization

Fluorescent *in situ* hybridization (FISH) is one of the nucleic acids techniques useful for studying microorganisms in their natural environments. The technique, first applied by DeLong *et al.* (1989), uses the ability of dye-labeled oligonucleotide probes to selectively bind to the bacterial ribosome. The dyed cells can then be counted with epiûuorescence microscopy, confocal laser scanning microscopy (CLSM), or ûow cytometry. Application of these procedures with multiple oligonucleotide probes can provide a general picture of microbial diversity (Rogers *et al.*, 2007). Alternatively, the use of highly specific probes such as 16S rDNA provides information on one or several species. Probes are available in hierarchical sets from the domain level (Bacteria and Archaea) to the species level based on the 16S rDNA database. Microautoradiography (MAR) is a process by which radioactive materials taken up by cells are visualized by exposure of a thin ûlm of photographic emulsion on a microscope slide. The exposed emulsion creates a pattern of silver grains surrounding the cells that can be seen clearly by transmission light microscopy. The basic methodology for FISH-MAR in soils and sediments can be divided into five steps including incubation with radioactive substrates, fixation and extraction, FISH, MAR, and microscopy (visualization).

The FISH-MAR technique can be used to identify the general metabolic activity of a microbial community, to study the *in situ* physiology of a defined (probe-specific) group of microorganisms, or to identify and quantify microorganisms responsible for a physiological process of interest in the environment

The application of FISH to soils and sediments is relatively recent compared with its use in marine and fresh water environments. Studies of community structure in soils have used oligonucleotide probes that target relatively large groups of bacteria of phylogenetically related genera. The most dominant groups in a subsurface arctic tundra soil were found to be of Beta-proteobacteria and the Cytophaga–Flavobacterium. The technique has also been successfully used to study influence of herbicides and urea on microbial community shift in the soil. Apart from this, FISH has also been used to examine rhizosphere (narrow zone around the root surface that is under the influence the root secretions) populations of various microorganisms (Watt *et al.*, 2006).

Stable isotope probing

Stable-isotope probing (SIP), a term coined to describe the use of ^{13}C-labelled growth substrates to link microbial function with identity via selective recovery of 'heavy' ^{13}C-labelled DNA (Radajewski *et al.*, 2000). In a wider context, however, SIP is not exclusively a DNA based technique. Analysis of labelled biomarkers including RNA and phospholipid-derived fatty acids (PLFAs) can also be used to identify the microorganisms involved in a particular function within a complex environmental sample.

The concept of DNA-SIP lies in the density difference of 'heavy' isotopically labeled DNA and 'light' unlabeled DNA, which are separated by density gradient centrifugation (*e.g.* caesium chloride or caesium trifluoroacetate) (Fig. 4). In brief, SIP involves the addition of a compound (typically a carbon substrate) labeled with a stable isotope. These labeled compounds can be added directly, as ^{13}C-labelled compounds, or assimilation of root exudates can be studied by growing plants in the presence of $^{13}CO_2$. Cell components of organisms that assimilate this substrate will be 'heavier', owing to the incorporation of the stable isotope. 'Light' and 'heavy' cell components can then be analyzed separately to distinguish 'total' and 'active' members of the community. Ideally, the labeled and unlabeled DNA from two clearly distinguished bands is clearly visible by UV illumination in the presence of ethidium bromide. The high sensitivity of PCR is sufficient to detect low amounts of labeled DNA by PCR-based molecular analysis (SSCP, TGGE and DGGE).

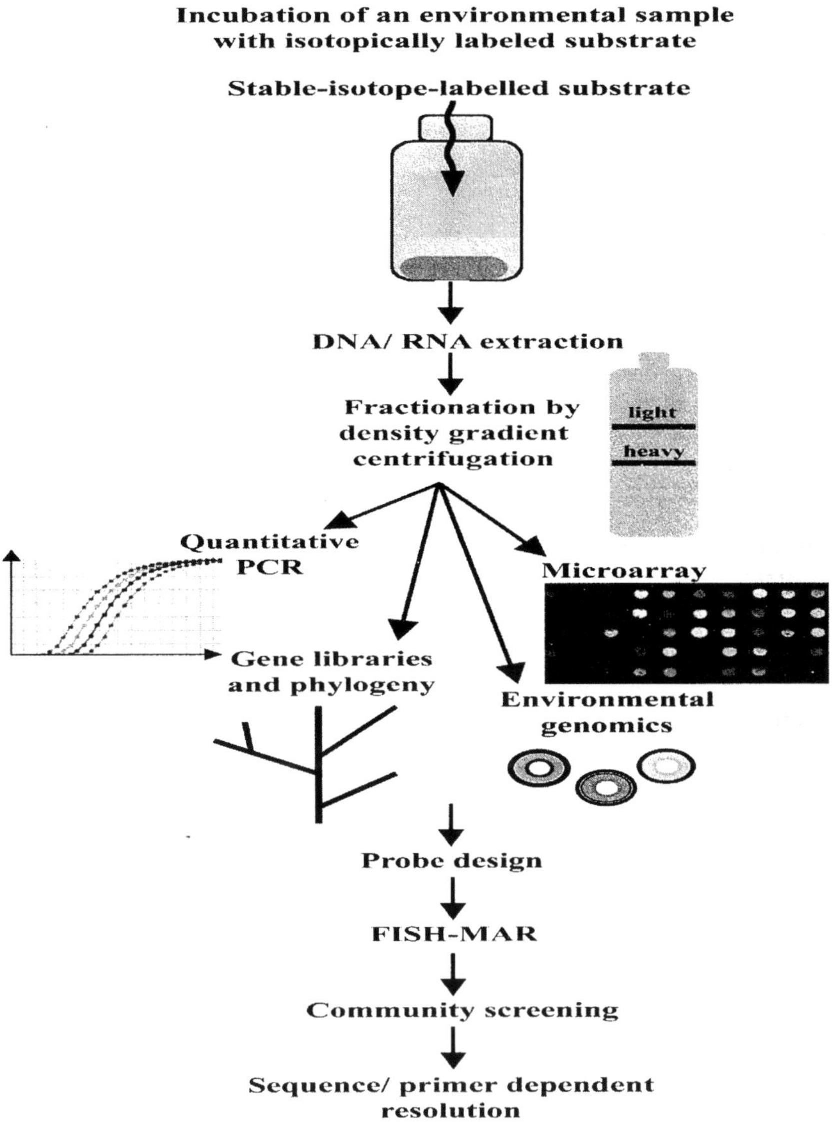

Fig. 4: Overview of the protocol of the Stable Isotope Probing approach (Wagner *et al.,* 2006)

DNA-SIP is dependent upon the availability of compounds that are highly enriched with a rare stable isotope. To date, the substrates $^{13}CH_3OH$, $^{13}CH_4$ or $^{13}CO_2$ (99 atom% ^{13}C; *i.e.* 99% of the carbon atoms are the ^{13}C isotope) have been added to soil samples or enrichment cultures for DNA-SIP. The most important feature of DNA-SIP is that the heaviest DNA (^{13}C-DNA) fraction collected following density-gradient centrifugation contains the combined genomes of a microbial population that is able to incorporate the labelled substrate into their nucleic acids. Therefore, 'universal' polymerase chain reaction (PCR) primers that amplify the SSU rRNA genes of Bacteria, Archaea or Eukarya are integral for the identification, *a priori*, of microorganisms involved in the process of interest. The functional genes of specific metabolic pathways can also study by designing specific primers (*e.g. mxaF*, which encodes the active site subunit of methanol dehydrogenase).

Like DNA-SIP, RNA SIP has also been very widely used for the study of microbial diversity of natural environment including soil. The principles of density-gradient centrifugation have also been used to link function with taxonomic identity using a very promising RNA-based stable-isotope probing approach (RNA-SIP). Manefield *et al.* (2002) reasoned that RNA could serve as a more responsive biomarker than DNA for use in SIP, because in active cells, RNA synthesis occurs at a high rate and labeling can also occur without the need for DNA synthesis or replication of the organism. The sequence-based resolution offered by SSU rRNA should permit identification of the microorganisms involved in a biological process, without any prior knowledge of their identity.

A fundamental difference between RNA-SIP and its DNA counterpart is that it relies on the 'transcriptome' of the active target microorganism(s). Therefore, one key advantage of the RNA-SIP approach is the natural amplification of the phylogenetic signature molecule (rRNA) in active cells. The development of stable isotope probing (SIP) has made this possible in many ecological and biotechnological contexts. RNA-based SIP technologies represent a significant leap forward for culture independent 'functional phylogeny' analyses, where specific consumption of a given compound carrying a ^{13}C signature can be associated with the small subunit ribosomal RNA molecules of the microbes that consume it.

To date, SIP has been carried out mostly in microcosms involving possible 'bottle effects' (*i.e.* possible artifacts of enriching certain microbial populations owing to unnatural incubation conditions in the laboratory). DNA-SIP is also feasible directly in the field by probing aerobic naphthalene degrading bacteria with small doses (22 ppm) of $^{13}C_6$-labeled naphthalene added directly to a coal tar waste contaminated aquifer. A predominant beta-proteobacterial *Polaromonas* sp. was detected by probing ^{13}C-labeled DNA by PCR, terminal restriction fragment length polymorphism (T-RFLP), and cloning of 16S rRNA genes, which has been described as *Polaromonas naphthalenivorans* only recently. Moreover, pulse labelling of photosynthesizing plants with $^{13}CO_2$ is the most direct approach to determine which organisms utilize plant root exudates. Pulse labeling of rice plants with $^{13}CO_2$ resulted in incorporation of ^{13}C into the ribosomal RNA of rice and a separate archaeal group has been identified suggesting that this particular group plays a key role in CH_4 production from plant-derived carbon. (Yahai and Conrad, 2005).

Metaproteomics

The term "metaproteomics" was first proposed by Wilmes and Bond (2004) as "the large-scale characterization of the entire protein complement of environmental microbiota at a given point in time". As proteins, and more precisely enzymes, are involved in biotransformation processes, metaproteome analysis constitutes a suitable way to characterize the dynamics of microbial function in a holistic way, which represents the last and crucial step toward our understanding of metabolome regulation (collective metabolites from all microorganisms present in an ecosystem). These studies provide insight into the functional dimensions of environmental genomic datasets and will help to achieve a major goal of environmental microbiology: the ability to link individual microbial species to function. Metaproteome analysis also provides physiological responses to changes in various environmental conditions.

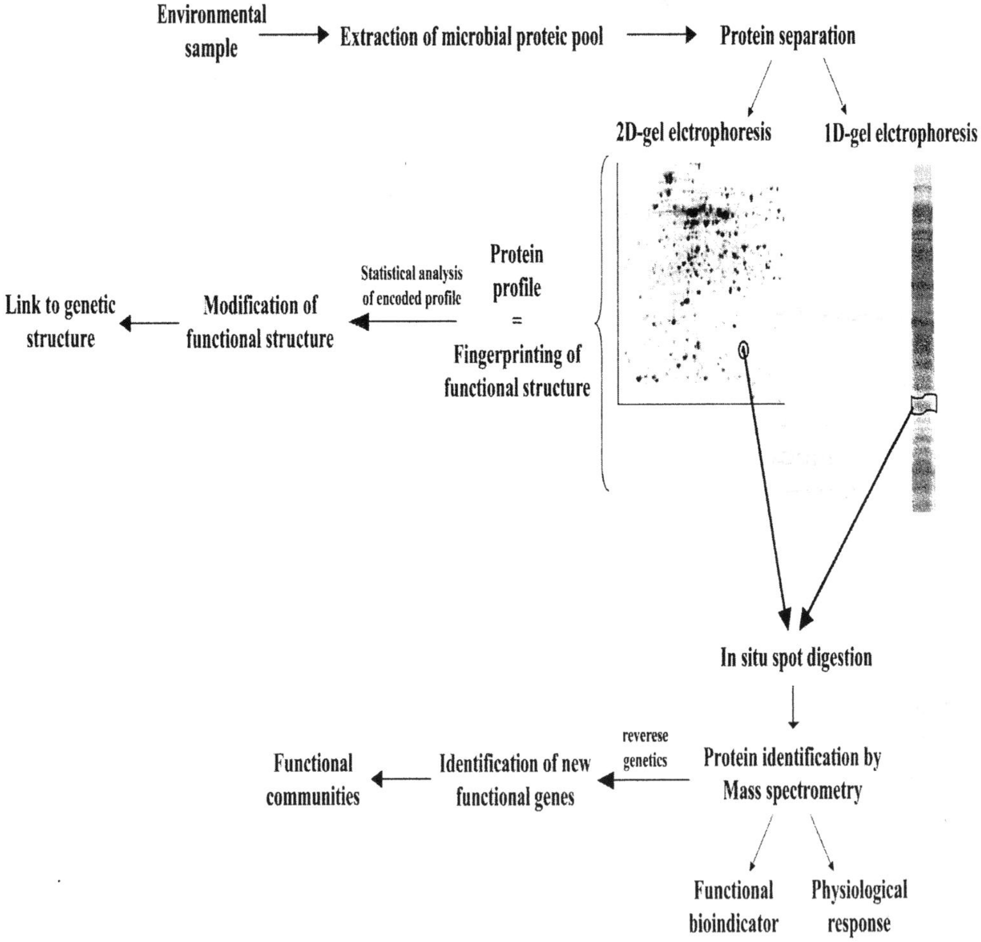

Fig. 5: Experimental strategy for metaproteome characterization (Maron *et al.*, 2007)

Metaproteome analysis of soil microbial communities implies the development of different technical steps, from the extraction of microbial proteins from the environmental matrix to the resolution of their diversity and their identification (Fig. 5). The purified protein sample is analyzed according to the type of information and the level of resolution required. **'Proteofingerprints'** of bacterial communities are obtained by one-dimensional and two-dimensional (2-D) gel electrophoresis, which is a technique used to separate proteins by both their isoelectric point (pI; in the first dimension) and by their mass (second dimension). The protein sample is first applied to a gel where an electric current is used to separate proteins by pI. After further processing, the same gel is applied to the top of second gel (a traditional SDS-PAGE gel), where application of an electric current will force the migration of the proteins from the first to the second dimension gel and the separation of protein occur based on their mass. 2-D gel electrophoresis is preferred to obtain a better protein separation.

The coupling of 2-D gel electrophoresis with mass spectrometry (MS) facilitates the further identification of a particular protein. The separated proteins of 2-D gel electrophoresis are revealed as distinct separate spots on the gel. These spots are removed and further analyzed by mass spectrometry to determine the identity of each individual protein spot.

The major limitation of the techniques is to extract pure protein fraction from soil sample because of the complexity of indigenous microbial communities, the heterogeneity of natural environment and also the presence of interfering compounds (phenolics and humic acids). Because of these limitations, metaproteomic approach has still not been used very frequently.

Statistical approaches to measure soil diversity

The sampling of natural microbial diversity from soil ecosystem has an inherent dilemma as to how efficient is the sampling procedure and reflects a community's "true" diversity. The huge data of microbial diversity are generated by new genetic techniques which explore large collection of hidden microbial world. A variety of statistical approaches have been developed to compare and estimate species diversity, richness and evenness from samples of microorganisms.

In ecology, these **diversity indices** are considered as significant *statistical tools* which are intended to measure the biodiversity of an ecosystem. More generally, diversity indices can be used to assess the diversity of any population in which each member belongs to a unique species. Estimators for diversity indices are likely to be biased, so caution is advisable when comparing similar values. Analyzing genetic diversity data using diversity indices is based on the banding pattern obtained. If all the DNA fragments were similar in the gel, they would find matching partners very quickly. But the more diverse the DNA strands, the longer this match-making takes, which gives an estimation of how many different species are there in the sample.

Microbial diversity of soil ecosystem can be quantified in many different ways. The two main factors taken into account when measuring diversity are richness and evenness. Richness is a measure of the number of different kinds of organisms present in a particular area. For example, species richness is the number of different species present. However, diversity depends not only on richness, but also on evenness. Evenness compares the similarity of the population size of each of the species present.

Richness

The number of species per sample is a measure of richness. Species richness is a measure of diversity that indicates the number of different species or different types of microorganisms that are present in a sample. In other terms, "the more are species present in a sample, the 'richer' is the sample". Species richness is a measure on its own takes no account of the number of individuals of each species present. It rather gives emphasis on the presence of different kind of individuals. It gives as much weight to those species, which have very few individuals as to those, which have many individuals.

Two well-known richness indices are Margalef (R1) (Margalef, 1958) and Menhinick (R2) (Menhinick, 1964) indices. R_1 is calculated as follows: $R_1 = (S - 1)/\ln(n)$, where S is the total number of species in a community and n is the total number of samples observed. R_2 (Menhinick index) is calculated as follows: $R_2 = S/$"n. D_1 and D_2 are sensitive to variations in sample size.

Evenness

Evenness is a measure of the relative abundance of the different species making up the richness of an area. When all species of a community are equally abundant, an evenness index should be maximum and decreased towards zero as the relative abundance of the species diverge away from evenness. The most common evenness index used by ecologists is E1 [H'/ln(S)].

A community dominated by one or two species is considered to be less diverse than one in which several different species have a similar abundance.

Diversity indices

Diversity indices incorporate both species richness and evenness into a single value. As species richness and evenness increase, so does diversity. Shannon's diversity index is a measure of diversity which takes into account both richness and evenness. The two most popular and widely used indices of diversity measurement are Shannon's diversity index and Simpson's diversity index.

Shannon's diversity index

Shannon's diversity index is the measure of the entropy (average degree of uncertainty) of a system and is named after ecologist Claude Shannon. The average uncertainty increases with the increase in the number of species and also as the distribution of individual among the species becomes even. The Shannon index (H') has probably been the most widely used index in community ecology. Two properties of H' made it a popular measure of species diversity; (i) H' = 0 if and only if there is one species in the sample, and (ii) H' is maximum only when all species (S) are represented by the same number of individuals, *i.e.*, a perfectly even distribution of abundances.

$$H' = \sum_{i=1}^{S^*} (p_i \ln p_i)$$

where H' is the average uncertainty per species in an infinite community made up of S^* species with known proportional abundances $p_1, p_2, p_3 \ldots . p_{S^*}$. p_i is calculated as n_i/n where n_i is the total number of organisms of a particular species and n is the total number of organisms of all species.

However, sometimes, there is also biasedness in the Shannon's diversity index measurement, because the total number of species in the community (S^*) will most likely be greater than the number of species observed in the sample (S). Fortunately, if n is large, this bias will be small (as found in most of the cases).

Simpson's Diversity Index

Simpson's Index (λ) measures the probability that two individuals randomly selected from a sample will belong to the same species (or some category other than species). There are two versions of the formula for calculating D and either is acceptable.

$$\lambda = \sum_{i=1}^{S^*} p_i^2$$

Where p_i is the proportional abundance of the *i*th species, which is given by:

$$\lambda = \sum_{i=1}^{S^*} (n_i / n)^2$$

Where, n_i is the total number of organisms of a particular species, and n is the total number of organisms of all species. The value of λ varies from 0 to 1. With this index, 0 represents infinite diversity and 1, no diversity. That is, "the bigger the value of λ, the lower the diversity". This is neither intuitive nor logical, so to get over this problem, λ is often subtracted from 1 to give (1–λ). The value of this index also ranges between 0 and 1, but here, "the greater the value, the greater the sample diversity". This makes more sense. In this case, the index represents the probability that two individuals randomly selected from a sample will belong to different species.

References

Cho-Smith, L.P., Maquelin, K., vanVreeswijk, T., Bruining, H.A., Pupples, G.J., Ngo Thi, N.A., Kirschner, C., Naumann, D., Ami, D., Villa, A.M., Orsini, F., Allocuh, P. and Endtz, H.P. 2001. Investigating microbial (micro) colony heterogeneity by vibrational spectroscopy. *Appl. Environ. Microbiol.*, **67**: 1461-1469.

DeLong, E.F., Wickham, G.S. and Pace, N.R. 1989. Phylogenetic stains: Ribosomal RNA-based probes for the identification of single microbial cells. *Science*, **243**: 1360-1363.

Dunbar, J., Ticknor, L.O. and Kuske, C.R. 2000. Assessment of microbial diversity in four southwestern United States soil by 16S rRNA gene terminal restriction fragment analysis. *Appl. Environ. Microbiol.*, **66**: 2943-2950.

Hurt,R.A., Qiu, X.Y., Wu, L.Y., Roh, Y.A., Palumbo, V., Tiedje, J.M. and Zhou, J.H. 2001. Simultaneous recovery of RNA and DNA from soils and sediments. *Appl. Environ Microbiol.*, **67**: 4495-4503.

Kennedy, A.C. 1994. Carbon utilization and fatty acid profiles for characterization of bacteria. In: Methods of soil analysis Part 2. Microbiological and biochemical properties of soil. (eds.) R.W.Waever, S.Angle, P.Bottomley, D.Bezdicek, S.Smith, A.Tabatabai and A.Wollum, Sci. Soc. America. Inc., Madison, Wis. pp. 453-553.

Liu, K.T. and Stahl, D.A. 2000. Molecular approaches for the measurement of density, diversity and phylogeny. In: Manual of Environmental Microbiology. (eds.) C.J.Hurst, R.L. G.R.Crawford, M.J. Knudsen, McInerney and L.D. Stetzenbach. ASM Press, Washington, USA. pp. 114-134.

Louws, F.J., Rademaker, J.L.W. and de Bruijn, F.J. 1999. The three Ds of PCR-based genomic analysis of phytobacteria: Diversity: Detection, and Disease diagnosis. *Ann. Rev. Phytopathol.*, **3**: 81-125.

MacNeil, I.A., Tiong, C.L., Minor, C., August, P.R., Grossman, T.H., Loiacono, K.A., Lynch, B.A., Phillips, T., Narula, S., Sundaramoorthi, R. *et al.* 2001. Expression and isolation of antimicrobial small molecules from soil DNA libraries. *J. Mol. Microbiol. Biotech.*, **3**: 301-308.

Manefield, M., Whiteley, A.S., Ostle, N., Ineson, P. and Bailey, M.J. 2002. Technical considerations for RNAbased stable isotope probing: an approach to associating microbial diversity with microbial community function. *Rapid Commun. Mass Spectrom.*, **16**: 2179-2183.

Margalef, R. 1958. Information theory in ecology. *Gen. Syst.*, **3:** 36-71.

Menhinick, E.P. 1964. A comparison of some species-individual diversity indices applies to samples of field insects. *Ecology,* **45**: 859-861.

MIDI. 1999. Sherlock Microbial Identification System software. Version 3.0. MIDI, Newark, DE.

Muyzer, G., Dewall, E.C. and Uitterlinden, A.G. 1993. Profiling of complex bacterial population by denaturing gradient gel electrophoresis analysis of polymerase chain reaction-amplified genes coding for 16S rRNA. *Appl. Environ. Microbiol.,* **59**: 695-700.

Pace, N.R., Stahl, D.A., Lane, D.J. and Olsen, G.J. 1985. Analyzing natural microbial populations by rRNA sequences. *ASM News,* **51**: 4-12.

Pace, N.R., Stahl, D.A., Lane, D.J. and Olsen. 1986. The use of 16S rDNA sequences to characterize natural microbial population. *Adv. Microb. Ecol.,* **9**: 1-55.

Peplies, J., Glõckner, F.O. and Amann, R. 2003. Optimization strategies for DNA Microarray-based detection of bacteria with 16S rDNA-targeting oligonucleotide probes. *Appl. Environ. Microbiol.,* **69**: 1397-1407.

Radajewski, S., Ineson, P., Parekh, N.R. and Murrell, J.C. 2000. Stable-isotope probing as a tool in microbial ecology. *Nature,* **403**: 646-649.

Riesenfeld, C.S., Schloss, P.D. and Handelsman, J. 2004. Metagenomics: Genomic Analysis of Microbial Communities. *Annu. Rev. Genet.,* **38**: 525-52.

Rogers, S.W., Moorman, T.B. and Ong, S.K. 2007 Fluorescent in situ hybridization and microautoradiography applied to ecophysiology in soil. *Soil Sci. Soc. Am. J.,* **71**: 620-631.

Schmidt, T.M., DeLong, E.F. and Pace, N.R. 1991. Analysis of a marine picoplankton community by 16S rRNA gene cloning and sequencing. *J. Bacteriol.,* **173**: 4371-78.

Schwieger, F. and Tebbe, C.C. 2000. Effect of field inoculation with *Sinorhizobium meliloti* L33 on the composition of bacterial communities in rhizosphere of a target plant (*Medicago sativa*) and a non target plant (*Chenopodium album*)-linking of 16S rRNA Gene based single strand conformation polymorphisms community profiles to the diversity of cultivated bacteria. *Appl. Environ. Microbiol.,* **66**: 3556-3565.

Schmalenberger, A. and Tebbe, C.C. 2002. Bacterial community composition in the rhizosphere of a transgenic herbicide-resistant maize (*Zea mays*) and comparison to its non-transgenic cultivar. *FEMS Microbiol. Ecol.*, **40**: 29-37.

Staley, J.T. and Kanopka, A. 1985. Measurement of *in situ* activities of nonphotosynthetic microorganisms in aquatic and terrestrial habitats. *Ann Rev. Microbiol.*, **39**: 321-346.

Tiedje, J.M., Chao, J.C., Murray, A., Treves, D., Xia, B. and Xhou, J. 2001. Soil teeming with life: new frontier for soil science, In: Sustainable management of soil organic matter (eds.) R.M.Rees, B.C. Ball, C.D.Campbell and C.A.Watson). CAB International, Wallingford, pp. 393-412.

Torsvik, V., Srheim, R. and Goksyr, J. 1996. Total bacterial diversity in soil and sediment communities – a review. *J. Ind. Microbiol.*, **17**: 170-178.

Versalovic, J., Schneider, M., deBruijn, F.J. and Lupski, J.R. 1994. Genomic fingerprinting of bacteria using repetitive sequence based polymerase chain reaction. *Methods Mol. Cell Soil,* **5**: 25-40.

Wagner, M., Nielsen, P.H., Loy, A., Nielsen, J.L. and Daims, H. 2006. Linking microbial community structure with function: fluorescence *in situ* hybridization-microautoradiography and isotope arrays. *Curr. Opin. Biotechnol.*, **17**: 83-91.

Ward, D.M., Ferris, M.J., Nold, S.C. and Bateson, M.M. 1998. A natural view of microbial biodiversity within hot spring cyanobacterial mat communities. *Microbiol. Mol. Biol. Rev.*, **62**: 1353-1370.

Watt, M., Hugenholtz, P., White, R. and Vinal, K. 2006. Numbers and locations of native bacteria on field grown wheat roots quantified by fluorescence *in situ* hybridization. *Environ. Microbiol.*, **8**: 871-884.

Wilmes, P. and Bond, P.L. 2004. The application of two-dimensional polyacrylamide gel electrophoresis and downstream analyses to a mixed community of prokaryotic microorganisms. *Environ. Microbiol.*, **6**: 911-920.

Woese, C.R. 1987. Bacterial Evolution. *Microbiol. Rev.*, **51**: 221-271.

Woese, C.R. and Fox, G.E. 1977. Phylogenetic structure of the prokaryotic domain: the primary kingdom. *Proc. Natl. Acad. Sci. USA.*, **74**: 5088-5090.

Yahai, L. and Conrad, R. 2005. *In situ* stable isotope probing of methanogenic Archaea in the rice rhizosphere. *Science,* **309**: 1088-1090.

Zumstein, E., Moletta, R. and Godon, J.J. 2000. Examination of two years of community dynamics in an anaerobic bioreactor using fluorescence polymerase chain reaction (PCR) single strand conformation polymorphisms analysis. *Environ. Microbiol,* **2**: 69-78.

Soil Microbiology and Soil Biotechnology : Vol. 1
Editor : S. Kannaiyan
Associated Publishing Co., New Delhi, 2010, pp. 158-167

11

Industrial Potentials of Microalgal Pigments

G.A. Ravishankar, R. Sarada, B. Sandesh Kamath*

Introduction

Microalgae have been known for centuries for their utility as a source of food. However in last three decades their potential has been realized as a unique source of biochemical molecules of value in food, cosmetics, feed and pharmaceutical sectors. These include compounds such as phycobiliproteins, carotenoids, polysaccharides, vitamins, sterols and bioactive molecules of different nature and fatty acids *viz.* gamma linolenic acids, eicosapentaenoic acid, docosahexaenoic acid. The exploitation of microalgae for nutritional purpose was limited to few organisms like *Spirulina*, *Chlorella*, *Scenedesmus* and β-carotene from *Dunaliella*. The main constraints in non-exploitation for mass cultivation of microalgae are their slow growth rate, low biomass yield and inadequate culture methodologies. With improvements in the cultivation methodologies and use of closed photobioreactors of different configuration *viz.* fermentation reactors, open reactors, tubular reactor have opened up new vistas for development of microalgal biotechnology (Fig. 1). The application of microalgae in aquaculture, food, speciality chemicals and environment has been the focus of microalgal research in recent times.

Of the diverse compounds produced by microalgae, the algal pigments such as phycobiliproteins and carotenoids have gained commercial importance owing to their unique and multiple applications as food and cosmetic colourants, antioxidants and several beneficial effects of nutraceutical and pharmaceutical value (Table 1).

Phycobiliproteins

Phycobiliproteins are the major accessory light harvesting pigments present in Cyanophyta (blue green algae), Cryptophyta (cryptomonad algae) and Rhodophyta (red algae). (1) The phycobiliproteins consists of protein moiety and linear tetrapyrrole prosthetic group (phycobilins) attached to the polypeptide through thioether bonds. Based on spectroscopic properties these proteins are broadly classified into 3 groups-allophycocyanin (650-655 nm), C-phycocyanin (610-620 nm) and R-phycoerythrin. (540-570 nm), each of these phycobiliproteins are comprised of 2 sub units α and β to which linear tetrapyrrole are

* Plant Cell Biotechnology Department, Central Food Technological Research Institute, Mysore - 570 020, India

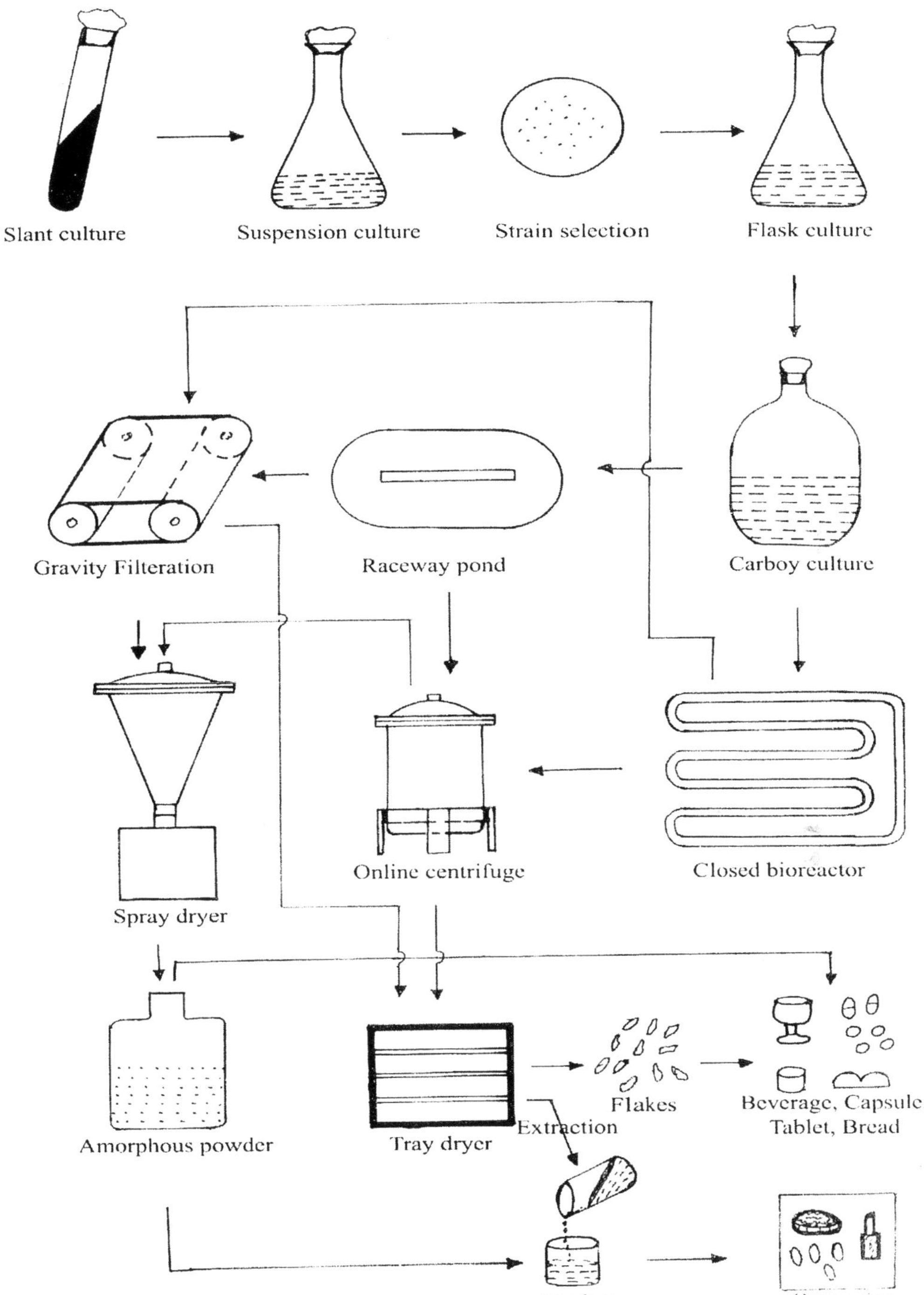

Fig. 1: Schematic representation of algal production process

covalently attached by a cysteine thioether bond. The central portion of all the 3 groups is an open tetrapyrrole containing skeletal system similar to chlorophyll and haemoglobin. Four types of tetrapyrroles are known–phycocyanobilin, phycoerythrobilin, phycourobilin and phycoerythrocyanin. These chromophoretic proteins along with some of the colourless linker peptides form an organized structure called phycobilisome (2). These phycobilisomes are situated on thylakoid membranes and channels the absorbed light energy to photosynthetic system II. Besides being light harvesting pigments, phycobiliproteins (C-phycocyanin) can also act as a nitrogen storage compound. Unlike other photosynthetic pigments, phycobiliproteins (PBP) are highly soluble in water, intensely coloured and fluorescent compounds. In purified form, they may exist as monomers (α, β) or diammers or trimers or hexamers or equilibrium mixtures of two or more of these aggregates depending on pH, ionic strength, solvent composition, protein concentration and temperature. The proteins are stable over a pH range of 5.0 to 9.0 (3)

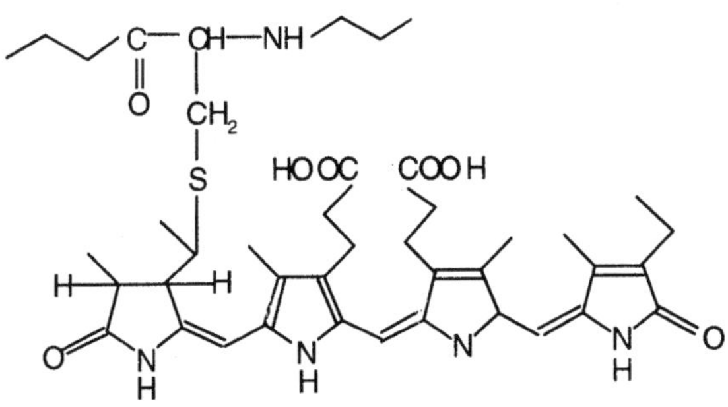

Structure of Phycocyanobilin

Phycobiliproteins owing to their intense colour and unusual characteristics of fluorescence in the visible range found application as fluorescent tag in immunodiagnostics besides its use as food and cosmetic colourant. Multinational companies like Dianippon Ink, Kaosoap, Ajinimoto etc. are marketing algal products for food additives at present. Japan seems to be forging ahead of all other countries in the use of these lesser known colourants. For example, Dianippon Ink. and Chemicals (Japan) has developed a blue food colour from *Spirulina* called "Lina blue" that constitutes 17% phycocyanin. Lina blue is used as a colorant in chewing gum, ice sherbets, frozen confectioneries, popsicles, candies, soft drink and dairy products. (4) Similarly Japanese industries use phycocyanin in eyeliners, eye shadows and lipstic formulations replacing the coal tar dyes. It is reported that this algal pigment has a significant market value of $ ~ 10 million per annum in Japan (www.phycobiliprotein.com). In Europe, the natural blue colour from *Spirulina* is already being sold replacing chemical blue colours (5).

These pigments are highly sensitive fluorescent dyes. The high absorbance coefficients over a wide spectral range, high quantum yield (Q = 0.3-0.9) and their fluorescence stability over wide pH range and non quenching of their emission by other biomolecules make them a highly suitable fluorescent markers (Phycoflour probes.). These markers can be easily linked to immunoglobulins, biotin, avidin and lectins. The Phycobiliprotein conjugates are excellent reagents for analysis of multiple cells surface markers by fluorescence activated cell sorting method. Since the emissions of phycoerythrin, phycocyanin and allophycocyanin can readily be distinguished, multiple phycofluors can be simultaneously used to quantify the amounts of several markers on cell surfaces. The price for purified phycobiliproteins range from US$ 3.00-17.00 per mg (www.phycobiliprotein.com) for specialized purposes. However commercial grade with 95% purity is around $1000/kg. They are 10-30 times more intense than conventional dyes and offer an alternative to unpopular radioactive markers. Besides its (phycocyanin) application as colorant and phycofluor probe, a study conducted in Japan on mice showed anti tumor effect of phycocyanin preparation. Another study suggested that phycocyanin improves the immune system by raising lymphocyte activity (6). A Japanese patent states that taking a small dosage of phycocyanin daily maintains or accelerates normal control cell functions that prevents generation of malignancy such as cancer or inhibits its growth or recurrence (5). Phycocyanin elicitor activity for secondary metabolite production in plant cell cultures was reported (7).

Table 1: Microalgal pigments and their uses

Algal pigments	Source	Uses	Reference
Carotenoids			
β-carotene	*Dunaliella salina*	Provitamin –A	35
		Food colour	36
Astaxanthin	*Haematococcus pluvialis*	Chicken feeds	37
		Salmon feeds	3, 38
Canthaxanthin	Some Chlorophytes	Colour	21
Lutein	Mutants of *Dunaliella*	Colour	
Other pigments			
Phycocyanin	*Spirulina*	Food colour	11
	Other blue green algae	Diagnostics	39
Phycoerythrin	*Porphyridium* and other red algae	Food colour	40
		Diagnostics	
Chlorophyll	*Spirulina* and green algae	Cosmetics	41

These phycobiliproteins from algae were found to have newer applications especially for therapeutic purpose. Reddy *et al.* (8) reported the selective inhibition of cyclooxygenase-2 which is reported to be an induced form of COX involved in prostaglandin biosynthesis. The selective inhibition of this COX-2 would exhibit anti-inflammatory analgesic effects without the deleterious side effects of non steroidal anti-inflammatory drugs. Vadiraja *et al.* (9) have reported inhibition of peroxyl radical induced lipid peroxidation in *in vivo* condition. Thus they established antioxidant property of phycocyanin on the basis of experiment carried out both *in vivo* and *in vitro.*

There are several patents now on phycocyanin preparation and formulations. In CFTRI a technology has been developed for mass cultivation of *Spirulina* in which phycocyanin comprises up to 15-20% on dry weight basis. The product purity is over 90% and a recovery of over 80% phycocyanin from the biomass. The safety of food grade phycocyanin has been established (10, 11, 12, 13).

Carotenoids

The carotenoid composition of green algae is similar to that of higher plants; the main carotenoids being β-carotene, lutein, violaxathin, zeaxanthin and neoxanthin. The antioxidant properties of algal carotenoids and the role they play in preventing many pathologies linked to oxidative stress have been demonstrated by Okuzumi *et al.* (14). Among the 600 carotenoids occurring in nature, only few carotenoids have gained commercial importance such as lycopene, β-carotene, canthaxanthin, zeaxanthin, astaxanthin and lutein. Attempts are focused on the production of these commercially important carotenoids using suitable algal strains.

β-carotene

β-carotene is one of the valuable algal pigments because of their nutraceutical and pigmentation properties. β-carotene from the alga *Dunaliella salina* was the first high value algal product commercialized (15).

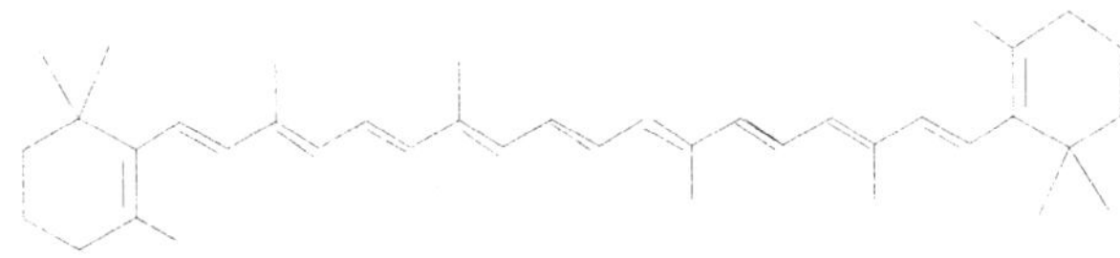

Structure of β-carotene

Dunaliella salina is the most halophilic eukaryote known, with an optimum salinity of about 22% w/v NaCl for growth, and an ability to withstand NaCl concentrations >30% and high temperatures (15). The cells contain upto 14% of its dry weight as β-carotene. The advantage of β–carotene from *Dunaliella* is that it consists of mixtures of the *cis* and *trans* isomers which are more biologically active than the synthetically produced pure *trans* b-carotene (16). Further more, the increasing evidence for a preventive action of β-carotene against certain cancers, particularly epithelial cancers (17) should lead to an increase in the demand for β-carotene.

Major producers of β-carotene are Western Biotechnology Ltd, Betatene Ltd in Australia, Microbio Resources Inc, Cyanotech in USA, Parry Agro in India and Naturebeta in Israel. The Australian producers use very large (5ha or greater) shallow ponds without any means of mixing (18), whereas the US, India and Israeli producers use paddle-wheel mixed raceway ponds to grow the algae. The β-carotene is sold mainly as an oil extract or as a suspension in vegetable oil, or as a dried algal powder (15). The β-carotene extracted in to vegetable oil can be applied as a food dye and a pro-vitamin additive for human consumption, for fish and poultry feed or in the cosmetics industry as an additive to sun-screen products. The price of this product will be comparatively lower and depends on the β-carotene concentration. The size of the market for natural β-carotene from *Dunaliella* is difficult to estimate but the trend and demand globally is increasing.

β-carotene have been shown to protect animals against UV-induced skin tumors, UV and carcinogen induced tumors (19). Similar observations have been made in cell and organ cultures where carotenoids have been shown to prevent malignant transformation and nuclear damage. Evidences have shown carotenoids may possess intrinsic chemo preventive action with respect to tumor formation.

Astaxanthin

Astaxanthin, a ketocarotenoid (3,3'-dihydroxy-β,β-carotene-4,4'-dione) is the principal pigment of salmonids and shrimps, various birds and many other organisms. Astaxanthin is a complex molecule and the synthesis, being difficult, results in an expensive product (US $ 2000-2500/Kg) (20). In the natural aquatic environment, astaxanthin is biosynthesized in the food chain within microalgae or phytoplankton at the primary production level. The microalgae are consumed by zooplankton, insects or crustaceans which accumulate astaxanthin and in turn, are ingested by salmonids. This requirement has created a market for astaxanthin worth about 100 million US dollars (21).

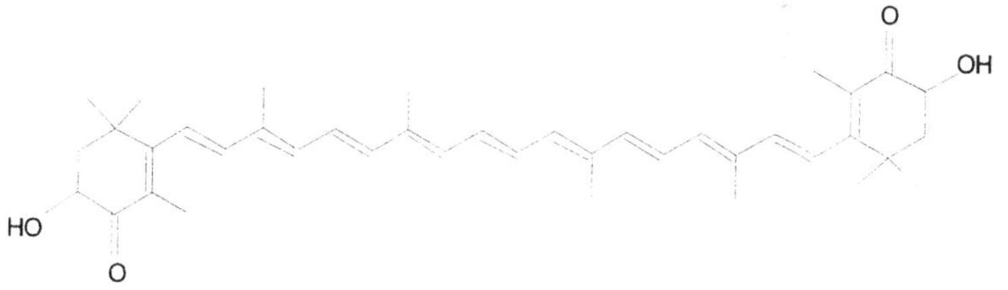

Structure of astaxanthin

Astaxanthin was first chemically identified by Kuhn and Sorenson (22). This molecule has two asymmetric carbons located at the 3 and 3' positions of the benzenoid rings on the other end of the molecule and can exist in four configurations which are present in natural sources of astaxanthin. Free astaxanthin and its mono and diesters from *Haematococcus pluvialis* have optically pure (3S, 3'S) chirality, whereas it is (3R, 3'R) from *Phaffia rhodozyma* (23).

Green alga *Haematococcus pluvialis*, is one of the potent natural sources of production of astaxanthin which constitutes 2-3% on dry weight basis. *Haematococcus* has two distinct phases in the life cycle- the motile vegetative phase and non motile encysted phase which accumulates carotenoid (24, 25). Stress conditions such as nutrient limitation (26), salinity stress (27), high light (28) induces astaxanthin accumulation which constitutes 85-90% of total carotenoid. The carotenoid exists as mono ester, diester and free form.

Astaxanthin being powerful bioactive antioxidant and has demonstrated efficacy in animal or human models of macular degeneration (29), a cause of blindness in the U.S., Alzheimer's and Parkinson's diseases, ameliorates the effects of LDL, protects against cancer and repairs the damage caused by lack of oxygen. Astaxanthin has been shown to be a powerful quencher of singlet oxygen activity in *in vitro* studies (30, 31). Astaxanthin has stronger antioxidant activity (10 times higher than β-carotene) and more than 500 times effective than α - tocopherol and astaxanthin has been proposed to be the super vitamin E (32). Mammals lack the ability to synthesize astaxanthin, or to convert dietary astaxanthin into vitamin A. Unlike β-carotene, astaxanthin has no pro-vitamin activity in these animals (33). Guerin *et al.* (34) reviewed the applications and possible role of astaxanthin in human health and nutrition.

Major producers of astaxanthin from microalgae–*Haematococcus pluvialis* are Hawaii-based firms, Cyanotech and Mera Pharmaceutical Inc.

There are several patents now on β–carotene and astaxanthin (Table 2). In CFTRI also efforts have been made to develop processes for the production of β–carotene and astaxanthin from *Dunaliella* and *Haematococcus* respectively. Few patents are also filed from the institute for cultivation method and downstream processing.

Table 2: Patents on microalgal pigments.

Title	Assignee	Patent No.
Medical uses of phycocyanin	The Regents of the University of California	US4886831
Coloured ophthalmic product	Thomas Hoxie Novartis Corporation	US 0033843
Modified phycobilisomes and the uses therefore	Brobeck, Phleger and Harrison LLP	US 0134325
An improved process for the preparation of a natural blue colourant phycocyanin from *Spirulina platensis.*	Council of Scientific and Industrial Research/Central Food Technological Research Institute	2504/Del/96
An improved process for the preservation of phycocyanin using synergistic formulation	Council of Scientific and Industrial Research/Central Food Technological Research Institute	367/Del/2000
High purity β-carotene	Natural Carotene Corporation U.S.	US 5310554
Carotenoid preparation	Israeli Biotechnology Research Ltd. Israel.	US 6383474
High *cis* β-carotene composition	Betatene Ltd. of Cheltenham, Australia	US 5612485
Method of obtaining a composition containing 9-*cis* β-carotene in high purity	Institute of Applied Biochemistry, Japan	US 6057484
Method for producing astaxanthin	F Hoffmann, La Roche	JP 202253266
Procedures for large scale production of astaxanthin from *Haematococcus*	Ben Gurion University of Negev Research and Development Authority	US 6022701
An improved process for the preparation of carotenoids from encysted *Haematococcus* cells.	Council of Scientific and Industrial Research/Central Food Technological Research Institute	1284/Del/1999
Process to produce astaxanthin from *Haematococcus* biomass	Parry Nutraceuticals Ltd	WO 03027267
An improved culture medium useful for carotenoid production	Council of Scientific and Industrial Research/Central Food Technological Research Institute	267/Del/2000

Conclusion

Thus the natural pigments from microalgae offers greater advantages than their synthetic counterparts. Although, synthetic pigments dominate the current market, there is an increasing awareness for natural pigments and their beneficial effects. Selection and content of desired compound in the organism, the cultivation method and the market would play a major role in the commercial success of the product. Improvements in the existing cultivation methods and newer cost effective devices would surely make the natural product cost competitive which may lead to its consumer acceptance. Phycocyanin from *Spirulina* is commercially available for food as well as diagnostic purpose. Similarly phycoerythrin from red algae is under exploitation. *Dunaliella* and *Haematococcus* are also cultivated in large scale as a rich source of carotenoid especially β-carotene and astaxanthin. Undoubtedly these algal forms are the richest source of these important dietary carotenoids essential for human physiology and disease prevention. Therefore, there is need for continued research from the biochemical, nutritional, nutraceutical, pharmaceutical and medical oriented angle.

References

Ben-Amotz, A. 1995. A new mode of *Dunaliella* biotechnology: two-phase growth for β-carotene production. *J. Appl. Phycol.*, **7**: 65-68.

Benemann, J.R. 1992. Microalgae aquaculture feed. *J. Appl. Phycol.*, **4**: 233-245.

Borowitzka, L.J. & Borowitzka, M.A. 1989. β-carotene (provitamin A) production with algae. In: Biotechnology of vitamins, pigments and growth factors, (Vandamme, E.J., ed). London: Elsevier Applied Science, pp. 15-26.

Borowitzka, L.J. & Borowitzka, M.A. 1990. Commercial production of β-carotene by *Dunaliella salina* in open ponds. *Bull. Mar. Sci.*, **47**: 244-252.

Boussiba, S. & Vonshak, A. 1991. Astaxanthin accumulation in the green alga *Haematococcus pluvialis. Plant. Cell. Physiol.*, **32**: 1077-1082.

Bryant, A.D., Guglielmi, G., de-Marsac, N.T. & Castets, A.M. 1979. The structure of cyanobacterial phycobilisomes: a model. *Arch Microbiol*, **123**: 113-127.

Bubrick, P. 1991. Production of astaxanthin from *Haematococcus. Biores. Technol.*, **38**: 237-239.

Dainippon Ink & Chemicals & Tokyo Kenkyukai. 1983. Antitumoral agents containing phycobilin- also used to treat ulcers and hemorrhoidal bleeding. JP 58065216 A 830418.

DiMascio, P., Devasagayam, T.P.A., Kaiser, S. & Sies, H. 1990 Carotenoids, tocopherols and thiols as biological singlet molecular oxygen quenchers. *Trans. Biochem. Soc.*, **18**: 1054-1056.

Francis, F.J. 1987. Lesser known food colorants. Food Technol., pp. 62-68.

Grung, M., D'Souza, M.L.D., Borowitzka, M.A. & Liaaen-Jensen, S. 1992. Algal carotenoids 51. Secondary carotenoids 2. *Haematococcus pluvialis* aplanospores as a source of (3S, 3'S) astaxanthin esters. *J. Appl. Phycol.*, **4**: 165-171.

Guerin, M., Huntley, M.E. & Olaizola, M. 2003. *Haematococcus* astaxanthin: applications for human health and nutrition.

Harker, M., Tsavalos, A.J. & Young, A.J. 1995. Use of response surface methodology to optimize carotenogenesis in the microalga *Haematococcus pluvialis. J. Appl. Phycol.*, **7**: 399-406.

Henrikson, R. 1989. New clinical research reveals health benefits. Earth food *Spirulina*. Ronore Enterprises, Inc. Laguna Beach, California. p. 69.

Johnson, E.A. & Schroeder, W.A. 1995. Microbial carotenoids. In: Advances in Biochemical Engineering and Biotechnology. (Fiechter A, Ed) Vol. 53, Springer-Verlag, Berlin, pp. 119-178.

Jyonouchi, H., Sun, S. and Gross, M. 1995. Effect of carotenoids on *in vitro* immunoglobulin production by human peripheral blood mononuclear cells: Astaxanthin, a carotenoid without vitamin A activity, enhances *in vitro* immunoglobulin production in response to a T-dependent stimulant and antigen. *Nutr. Cancer.*, **23**: 171-183.

Kobayashi, M., Kakizono, T., Yamaguchi, K. & Nagai, S. 1992. Effects of light intensity, light quality, and illumination cycle on astaxanthin formation in a green alga *Haematococcus pluvialis. J. Ferment. Bioeng.*, **74**: 61-63.

Krinsky, N.I. 1989. Carotenoids and cancer in animal models. *J. Nutr.*, **11**: 123-126.

Kuhn, R. & Sorenson, N.A. 1983. Uber astaxanthin and ovoerdin. *Ber. Dtsch. Chem. Ges.*, **71**: 1879.

Manoj, G., Raghava Rao, K.S.M.S., Jayaprakasan, S.G., Ravishankar, G.A., Ajithkumar, T. & Venkatraman, L.V. 1996. An improved process for the preparation of a natural blue colourant phycocyanin from *Spirulina platensis*. Indian patent, 2504/Del/96.

Manoj, G.P., Sarada, R. & Ravishankar, G.A. 2000. An improved process for the preservation of phycocyanin using synergistic formulation (367/Del/2000).

Miki, W. 1991. Biological functions and activities of animal carotenoids. *Pure. Appl. Chem.*, **63**:141-146.

Mokady, S., Avron, M. & Ben-Amortz. 1990. Accumulation in chick livers of 9-*cis* versus all-*trans* β-carotene. *J. Nutr.*, **120**: 889-892.

Naidu, A., Sarada, R., Manoj, G., Khan, M.Y., Mahadevaswamy, M., Vishwantha, S., Narasimha Murthy, K., Ravishankar, G.A. & Srinivas, L. 1999. Toxicity assessment of Phycocyanin: a blue colourant from blue green alga *Spirulina platensis. Food Biotechnol.*, **13**: 51-66.

Okuzumi, J., Tkahashi, T., Yamane, T., Kitao, Y., Inagake, M., Ohya, K., Nishino, H. & Tanaka, Y. 1993. Inhibitory effects of fucoxanthin, a natural carotenoid, pm N-ethyl-N[1]-nitro-N-nitrosoguanidine-induced mouse duodenal carcinogenesis. *Cancer Lett.*, **68**: 159-168.

Palozza, P. & Krinsky, N.I. 1992. Antioxidant effects of carotenoids in vivo and *in vitro*: An overview. (L Packer, ed). San Diego: Academic Press Inc.

Rao, R.S., Sarada, R. & Ravishankar, G.A. 1996. Phycocyanin, a new elicitor for capsaicin and anthocyanin accumulation in plant cell cultures. *Appl. Microbiol. Biotechnol.*, **46**: 619-621.

Reddy, M., Vadiraja, B.B., Kiranmai, G., Reddy, M.N., Reddanna & Madyastha, K.M. 2000. Selective inhibition of cyclooxygenase-2 by C-phycocyanin, a biliprotein from *Spirulina platensis. Biochem. Biophy. Res. Comm.*, **277**: 599-603.

Richmond, A. 1990. Large scale Microalgal culture and applications. In: Progress in Phycological Research (Round & Chapman eds) Vol 7. Biopress Ltd. pp. 1-62.

Sarada, R., Pillai, M.G & Ravishankar, G.A. 1999. Phycocyanin from *Spirulina* sp: influence of processing of biomass on phycocyanin yield, analysis of efficacy of extraction methods and stability studies on phycocyanin. *Process Biochem*, **34**: 795-801.

Sarada, R., Ramachandra Rao & Ravishankar, G.A. 1996. An improved method of effluxing of phycocyanin from cyanobacterium *Spirulina platensis* species. Indian patent, 1029/Del/1996.

Sarada, R., Sila Bhattacharya & Ravishankar, G.A. 2002. Optimization of culture conditions for growth of green alga *Haematococcus pluvialis. World J. Microbiol. Biotechnol.*, **18**: 517-521.

Sarada, R., Sila Bhattacharya, Suvendu Bhattacharya & Ravishankar, G.A. 2002. A response surface approach for the production of natural pigment astaxanthin from green alga

Haematococcus pluvialis: Effect of sodium acetate, culture age and sodium chloride. *Food. Biotechnol.*, **16**: 107-120.

Sarada, R., Usha, T. & Ravishankar, G.A. 2002. Influence of stress on astaxanthin production in *Haematococcus pluvialis* grown under different culture conditions. *Process. Biochem.*, **37**: 623-627.

Shaish, A., Ben-Amotz, & Avron, M. 1992. Biosynthesis of β-carotene in *Dunaliella. Meth. Enzymol.*, **213**: 439-444.

Snodderly, D.M. 1995. Evidence for protection against age-related macular degeneration by carotenoids and antioxidant vitamins. *Am. J. Clin. Nutr.*, **62**(suppl.): 1448S-1461S.

Stahelin, H.B., Gey, K.F., Eichhjolzer, M. & Ludin, E. 1991. β-carotene and cancer prevention—The Basel Study. *Am. J. Clin. Nutr.*, **53**: S265-S269.

Standnichuyk, I.N. 1995. Phycobiliproteins: determination of chromophore compostion and content. *Phytochem. Anal.*, **6**: 281-288.

Tandeau de Marsac, N. and Cohen-Bazire, G. 1977. *Proc. Natl. Acad. Sci.*, USA, **74**: 1635-1639.

Vadiraja, B.B. and Madyastha, K.M. 2000. C-Phycocyanin: A potent peroxyl radical scavenger *in vivo* and *in vitro. Biochem. Biophy. Res. Comm.*, **275**: 20-25.

Wilson, M.R., Crowley, S., Odgers, G.A. & Shaw, L. 1991. Immunofluoresecent labeling using covalently anti-phycoerythrin antibodies and phycoerythrin polymers. *Cytometry*, **12**: 373-377.

Soil Microbiology and Soil Biotechnology : Vol. 1
Editor : S. Kannaiyan
Associated Publishing Co., New Delhi, 2010, pp. 168-187

12

Trichoderma—a Potential Biotool for Beneficial Applications

S. Nakkeeran[1] and T. Marimuthu[2]

Introduction

Despite the huge diversity, the technological application for protecting the environment and also to promote health of plants, soils and human beings are underexploited. Fungi being the fifth kingdom of eukaryotic heterotrophic organism, out of 1.6 million species distributed world wide only about 1,50,000 species have so far been identified. Among several fungi, the genus *Trichoderma* is a predominant cosmopolitan fungus distributed in different soils of different agroclimatic zones. *Trichoderma* can grow rapidly on diversified substrates and can resist the noxious chemicals and the toxic heavy metals. The fungus with their well-developed enzyme system plays a vital role in improving both plant and soil health apart from its industrial applications. Besides, the antibiotics and lytic enzymes are used for the management of various plant pathogenic microbes infecting plants. Based on the efficacy of the genus *Trichoderma,* several products have been registered both at national and international levels.

Perception of the public reflects that the negative aspects of pesticides seem to outnumber their benefits. The increased reflection on environmental concern over pesticide use was instrumental in a large upsurge of bio-agents for both pest and disease management.

Development of fungicide resistance among the pathogens, ground water and foodstuff pollution and the development of oncogenic risks have further encouraged the exploitation of antagonistic microflora in disease management.

The decades of laboratory experiments have led to the explosion in the field of bio-control of fungi and introduction of more than 50 commercial formulations comprising fungi and bacteria (Whipps, 1997). Among the antagonists, *Trichoderma* plays a vital role in plant disease management and has opened new vistas for the commercialization. *Trichoderma* species are one of the predominant component among the mycoflora in various soils in all

1. Associate Professor & 2. Professor of Plant Pathology, Tamil Nadu Agricultural University, Coimbatore - 641 003

climatic zones. *Trichoderma* (Hypocreales, Ascomycota) species frequently associated with decaying wood produce more enzymes of industrial value (*Trichoderma reesei=Hypocrea jecorina*) (Kubicek and Penttila, 1998) and antibiotics (Sivasithamparam and Ghisalberti, 1998), or have application as biocontrol agents against plant pathogens (i.e. *T. harzianum=H. lixii*; *T. atroviride=H. atroviridis*; *T. asperellum*) (Hjeljord and Tronsmo, 1998). More recently, *T. longibrachiatum* was found to be an opportunistic pathogen of immuno compromised mammals including humans (Kredics *et al*., 2003). *Trichoderma* has been recognized to comprise a significant amount of fungal biomass in soil (Nelson, 1982) and is frequently present as an indoor contaminant (Thrane *et al*., 2001). These diverse implications of *Trichoderma/Hypocrea* with human society necessitate an accurate species and strain identification. However, classical approaches based on the use of morphological criteria are difficult to apply to *Trichoderma* due to the plasticity of characters. However, the knowledge on the diversity and distribution pattern of *Trichoderma* is highly essential to explore the best among the diverse species of *Trichoderma*. The analysis of the *Trichoderma* products registered with Central Insecticides Board (CIB), Faridabad, Haryana revealed that only *T. viride*, *T. harzianum* and *T. hamatum* have been explored to a maximum extent.

Eco-biology of *Trichoderma*

The genetic variation is a key factor to evaluate biodiversity. The genetic variations in *Trichoderma* are mainly attributed to their ability to adapt to different ecological conditions. *Trichoderma* occurs in all soils and other natural habitats that are rich in organic matter. *Trichoderma* is also found on root surfaces of various plants, decayed bark, mainly after the decay by other fungi (Elisa Esposito and Manuela da Silva, 1998). The population of *Trichoderma* decreases, when it is exposed to dry conditions for a long period of time. Some strains of *T. hamatum* and *T. pseudokoningii* are adapted to excessive soil moisture and those of *T. viride* and *T. polysporum* are found to be restricted to areas of low temperature. *T. harzianum* is predominant in warm climatic regions, while *T. hamatum* and *T. koningii* are widely distributed in areas of diverse climatic conditions (Papavizas, 1985; Grondona *et al*., 1997). It is also distributed in different soils including agricultural, forest, prairie, salt marsh and desert soils scattered in different agro-climatic zones. The propagules of *Trichoderma* were abundant in forest soils and they proliferated well in the forest soils rich in carbon and nitrogen (Danielson and Davey, 1973). Some of the species of *Trichoderma* survive as decomposers of woody and herbaceous materials and are also necrotrophic against the primary wood decomposers (Rossman, 1996).

The population dynamics of *Trichoderma* is also altered due to the variations in the iron content of the soil (Xu, 1996). Certain strains of *Trichoderma* have evolved and occupied the niches of higher elevations. Analysis of soil samples from higher elevations of about 2700m in the Himalayas reflected the presence of section *Longibrachyatum*. Based on the parsimony analysis section *Longibrachyatum* was grouped under *H. audinensis*. It was also isolated from 2300m elevation of Venezuelan Andes (Samuels *et al*.,1998). Another member of the section *Longibrachyatum* namely, *T. konilangbra* was isolated from locations between 1750 and 3400m in Uganda.

pH and *Trichoderma* diversity

Hydrogen ion concentration played a major role in the adaptability of biocontrol agents to their nutrient environment. The activities of *Trichoderma* spp. were more pronounced at low pH and the efficacy was influenced by acidic conditions (Chet and Baker,1980, 1981).The activity of *T. lignorum* was poor in neutral and alkaline soils (Dhingra and Khare,1973). Acidic pH favoured disease suppression induced by *T. harzianum* (Chet and Baker, 1980; Liu and Baker, 1980; Marshall, 1982; Harman and Taylor, 1988) and *T. hamatum* (Chet and Baker, 1981). The spore germination (Chet and Baker, 1980), mycelial growth, conidiophore production (Chet and Baker, 1980), antibiotic production (Dennis and Webster, 1971) or the activity of lytic enzymes (Chet and Baker, 1980) were enhanced under acidic conditions. Jeyarajan *et al.*(1994) found that the isolates of *T. viride* performed on a wide range of pH varying from 5.0 to 9.0. Since the isolates of *T. viride* perform in wide range of soil pH, a composite, compatible formulation of *T. viride* could easily be developed and used in disease management programs.

Moisture and *Trichoderma* diversity

The soil moisture content is very important than soil reactions in altering the population dynamics of the soil. It affects the growth of antagonist. The bacterial population dropped rapidly as the soil moisture was falling below 15 to 20 per cent while fungi and actinomycetes population were not affected. The minimum relative humidity required for the growth of *Trichoderma* is 91 per cent. The maximum growth of *Trichoderma* on ground pine needles occurred at 144 and 201 per cent moisture. Growth was good at 103 per cent but was depressed at 46 per cent moisture. Growth of *T. viride* was more at 40 to 60 per cent moisture than at higher soil moisture levels (Krishnamoorthy, 1987). Jeyarajan *et al.* (1994) observed that all the isolates of *T. viride* and *T. harzianum* survived uniformly well at 40 per cent followed by 60 per cent moisture holding capacity.

Ecology and the diversity of *Trichoderma*

The diversity of the species of *Trichoderma* varies based on locations and with the microsites. The soils of Taiwan were rich in *T. asperellum*. But, the tree barks associated with lichen in Chitan forest in Taiwan was colonized by *Hypocrea jecorina, T. viride, T. atroviride* and *T. harzianum.* On the contrary, the tree bark, which was not having association with any of the fauna was dominated by the association of *T. sp. novum.* Hence, based on the association of different fauna, the diversity of *Trichoderma* spp. also differs to a greater extent.

Similarly the tropical forest soil of Cambodia was rich in *T. harzianum*, which indicates that it prefers warm soil temperature for its survival and multiplication. Analyses of the samples from the ruined walls of the temple indicate that they were predominant with *T. harzianum* and to certain extent the ruined walls were also found to be associated with *T. virens*. The soils near seashore of Thailand at Coral Island and Koh Samet Island were colonized with *T.sp. novum* and *T. harzianum*. The wet soils of Kauchaburi province of Burma was colonized by *T. harzianum* and *T. ghanense*. The tropical rain forest soils of Malaysia had *T. virens, T. spirale, T. harzianum, T. asperellum* and *T. sp. novum.* The lake garden soil of Kuala Lumpur was associated with *T. virens, T. spirale, T. harzianum, T. sp. novum, T. koningii* and *T. atroviride*. The forest soil of Singapore was colonized with *T. koningii, T. asperellum, T. virens, T. spirale* and *T. harzianum*. Though there was a huge variation in the distribution pattern of the *Trichoderma* spp., the cultivated garden soils of South East Asia were predominant with *T. virens* which is being widely used as biocontrol fungi for the management of soil borne pathogens. This vividly explains that the ecological requirements are different for the species colonizing tree bark, forest soils and cultivated garden soils (Kubicek *et al.*, 2003).

Isolates of *T. spirale* were found to be more in South East Asia than the cosmopolitan *T. koningii*. Similarly several isolates of *T. spirale* were observed in soils of Canada. However, the abundance of the same was more in Central America, Columbia and South Africa which suggests that these species are more abundant in tropical climates (Kubicek *et al.*, 2003). *H. jecorina,* a pantropical ascomycete is abundant in South East Asia, the Pacific Island and South America. But the same was previously reported to be restricted to the regions around equator (Samuels *et al.*, 1998). Soils from Central Russia, The Urals, Siberia and the Himalayan mountains were predominant with *T. atroviride, T. hamatum, T. virens, T. asperellum, T. koningii, T.harzianum and T. oblongisporum* (Kullinig *et al.*, 2000).

Biodiversity of *Trichoderma*

Trichoderma being cosmopolitan fungus found to be widely distributed in different agroclimatic zones and habitats in both tropical and subtropical countries. However, the biogeographical distribution of *Trichoderma* has not been extensively studied except in European countries, Russia and China. Since *Trichoderma* has wide scope and potential to improve human life in different angles, attention must be focused to identify the diversity and distribution pattern of *Trichoderma*. About 35 *Trichoderma* species are recognised on the basis of morphological and molecular characters. The diversity of *Trichoderma* spp. in Russia, Siberia and the Himalayan mountains were identified through molecular approaches using ITS 1 and 2 sequences of the rDNA cluster of the 75 isolates obtained. The comparison of sequences confirmed the presence of *T. atroviride*, *T. virens*, *T. hamatum*, *T. asperellum*, *T. koningii* and *T. oblongisporum*, *T. harzianum* and *T. inhamatum*. Sections *Longibrachiatum* and *T. stromaticum* were found distributed at an elevation of 2700 m in the Himalayas (Kullnig *et al.*, 2000).

Parsimony and distance analyses of DNA sequences from multiple genetic loci of *Trichoderma* with 18S rDNA sequence analysis suggested that the genus *Trichoderma* might have evolved 110 million years ago as *Hypomyces* and *Fusarium*. 28S rDNA sequence analysis shows that the genus *Trichoderma* is part of a monophyletic branch within the Hypocreaceae. Phylogenic trees inferred from a combined analysis of the nuclear ribosomal internal transcribed spacer provided strong statistical support for a phylogeny consistent with the existence of four clades. Clade A comprises of species of Bissett's (1991 a,b,c) sect. *Trichoderma* in addition to *T. hamatum*, *T. pubescens*, *T. asperellum* and *T. strigosum*; clade C comprises all the species contained in section *Longibrachiatum* as revised by Samuels *et al.* (1998) and clade D contains only *T. aureoviride* which is genetically most distant to all other species. Clade B, on the other hand, contains a large and taxonomically heterogeneous mixture of species, among which several subclades could be identified: subclade B1 containing *H. lactea*, *H. citrina*, *H. citrina* var. *americana*, *H. lutea* and an unnamed *T.* sp. 1; subclade B2 containing *T. stromaticum* and an unnamed *T.* sp. PPRI3559; subclade B3 containing *T. fertile*, *T. oblongisporum* and *H. hunua*; subclade B5 containing *T. polysporum*, *T. roceum*, *Hypocrea pilulifera* and *T. minutisporum* and a larger subclade (B4), in which three strain clusters could be distinguished : one comprising *T. harzianum*, *T. inhamatum*, *H. vinosa* and *T. aggressivum*, another one containing *T. fasciculatum*, *T. longipile*, and *T. strictipile* and a third containing *T. virens*, *T. avofuscum*, and *T. crassum*. The position of the remaining species of subclade B4 (*T. spirale*, *T. cfr aureoviride*, *H. tawa*, and *T. tomentosum*) was not resolved (Kullnig Gradinger *et al.*, 2002).

The variations of *Trichoderma* in South East Asia (Taiwan and Indonesia) were identified through morphological, biochemical and through sequence analysis. Ninety six strains were isolated. Seventy eight isolates were identified as *T. harzianum/T. inhamatum* (37 strains), *T. virens* (16 strains), *T.spirale* (8 strains), *T. koningii* (3 strains), *T. atroviride* (3 strains), *T. asperellum* (4 strains), *Hypocrea jecorina* (anamorph: *T. reesei*; 2 strains), *T. viride* (2 strains), *T. hamatum* (1 strain) and *T. ghanense* (1 strain). Analysis of biochemical characters revealed that *T. virens, T. spirale, T. asperellum, T. koningii, H. jecorina,* and *T. ghanense* formed well defined clusters, thus exhibiting species specific metabolic properties. In addition, two new taxa of *Trichoderma* were identified (Kubicek *et al.*, 2003).

The biodiversity and biogeography of *Trichoderma* in four areas of China *viz.*, North (Hebei province), South-East (Zhejiang province), West (Himalayan, Tibet) and South-West (Yunnan province) include one hundred and thirty five isolates. They were grouped tentatively based on morphological characters which showed the existence of 64 isolates. But, all were grouped into eleven species as *Trichoderma asperellum, T. koningii, T. atroviride, T. viride, T. velutinum, T. cerinum, T. virens, T. harzianum, T. sinensis, T. citrinoviride, T. longibrachiatum* and two putative new species (*Trichoderma* sp.C1, and *Trichoderma* sp. C2) by both morphological characters and phylogenetic analysis. *T harzianum* accounted for almost half of the biodiversity and was predominant in North China. In addition, a unique haplotype of *T. harzianum* which is rarely found elsewhere was also identified (Zhang *et al.*, 2005).

Ecological significance of *Trichoderma*

Trichoderma plays an immense role in creating a pollution free environment through multifarious actions. *Trichoderma* could be used for:

- Plant growth promotion and disease control
- Improving soil health
- Bioremediation of polluted soils
- Industrial applications
- Phytobial remediation

Biotechnological applications

Biological control of plant pathogens

Trichoderma is the most widely used single genus as a potential biocontrol agent all over the world. *T. viride, T. harzianum, T. virens* and *T. hamatum* are the most exploited species for the management of various seed and soil borne pathogens of agricultural and horticultural crops (Table 1). Diseases due to *Rhizoctonia solani, Macrophomina phaseolina* and *Fusarium* spp. cause high percentage of mortality of the plants and consequent yield losses in rain-fed crops especially in pulses and oilseeds. Soil incorporation of *T. harzianum* in the field reduced the inoculum density of *R. solani* by 85 per cent and the fruit rot of tomato by 27-51 per cent. Soil application of *T. harzianum* was found effective in reducing the incidence of *Macrophomina* root rot of beans by 37 to 74 per cent. Coating of melon seeds with conidia of *T. harzianum* reduced the disease caused by *Macrophomina* to the extent of 46.3 per cent. Seed coating of green gram and cowpea @ 10g of *T. viride* isolate MG6 combined with soil application reduced dry root rot incidence and increased the yield (Nakkeeran and Sabitha Doraisamy, 2001). In addition to disease reduction, dry matter production was also enhanced which was attributed to the production of growth hormones. Treatment of sunflower seeds with *T. viride* @ 4g/kg reduced the root rot incidence to 18.15 as against 37.1 per cent in control. It increased the yield by 21 per cent. Root rot disease of blackgram could be managed by seed and soil application with *Trichoderma* (Dinakaran and Marimuthu, 2001).

The root rot incidence of *Sesamum* was reduced to 12.8 per cent when the seeds were treated with *T. viride* @ 4g/kg and increased the root, shoot length and oil content. It recorded an yield of 677 kg/ha as against 301 kg/ha in control. Results of the field demonstrations conducted in the farmer's fields for the control of root rot in groundnut, sesame, urdbean and chickpea against dry root rot through *T. viride* were found to be promising. *T. harzianum* was found effective in the management of *Fusarium* wilt diseases. Strain T35 checked *Fusarium* wilt of cotton and melon under natural soil conditions. Incorporation of *T. harzianum* as wheat bran-peat preparation into soil under green house conditions reduced the incidence of crown rot of tomato up to 80 per cent. Soil application increased the yield of tomato by 18.8 per cent.

Soil application of *T. koningii* @ 2 g per vine of black pepper at the time of transplanting was highly effective compared to metalaxyl compounds against *Phytophthora* foot rot. The disease incidence was 23 to 35 per cent lower and yield was 30 to 40 per cent higher than control. Pre and post planting application of *T. viride* and *T. harzianum* was effective against foot rot disease of betel vine. Amendment of nursery soils with *T. harzianum* prior to the colonization of damping off pathogen in nursery beds resulted in the reduction of pre and post emergence damping off of cardamom. Delivery of *Trichoderma* strains through solid matrix priming in tomato and chillies reduced the incidence of damping off under field conditions than dry seed treatments (Nakkeeran and Sabitha Doraisamy, 2001; Kannaiyan *et al.*, 2002). Bio-manure developed with coir pith and *T. viride* contained damping off disease of

Eucalyptus seedlings (Kumar and Marimuthu, 1993) and turmeric rhizome rot (Ushamalini *et al.*, 2006).

Table 1: *Trichoderma* species used against plant pathogens

Trichoderma species	Pathogen	Crop	Delivery system
Trichoderma sp. C62	*Sclerotium cepivorum*	Onion	Soil application
Trichoderma spp.	*G. graminis var. tritici R. solani*	Wheat Lettuce Lettuce	Spores added to soil Soil
T. harzianum	*Fusarium graminearum*	Radish	Soil
T. harzianum T95I	*Pythium ultimum*	Cucumber	Soil
T. koningii	*S. rolfsii*	Tomato	Soil
T. koningii	*Pythium* spp. *Pythium* spp.	Chickpea Pea	Seed Seed and Soil
T. harzianum	*P. ultimum*	Cotton Maize Pea	Seed Seed Seed
T. harzianum + *Rhizobium*	*S. rolfsii*	Groundnut	Seed and soil
T. viride, T. harzianum, T. longibrachiatum	*M. phaseolina*	*Sesamum*	Seed
T. viride	*M. phaseolina*	Groundnut *Sesamum* Blackgram Greengram Pigeon pea Sunflower	Seed
T. hamatum	*R. solani*	Cotton Sugar beet Radish	Soil-as mycelial preparations
T. viride	*M. phaseolina*	Blackgram	Seed
T. viride	*M. phaseolina*	Greengram	Furrow application
T. viride	*F. udum, M. phaseolina*	Pigeon pea	Seed and Soil
T. viride	*F. oxysporum f. sp.cubense* *R. solani* *M. phaseolina*	Banana Cotton Green gram Cowpea	*Trichoderma* capsules Seed Seed Seed + Soil
	Pythium ultimum	Chilies, Tomato	Solid Matrix priming
Trichoderma spp.	*M. phaseolina*	*Sesamum* Blackgram	Seed Seed and soil
T. viride	*R. solani* *S. rolfsii*	Cotton Brinjal Sunflower Bhendi	Seed

(Kannaiyan *et al.*, 2002)

Commercial production and formulations

It is the formulation that can transform a potential biological control agent from a laboratory curiosity to a commercially successful venture in terms of disease management (Connick *et al.*, 1990). A successful bio-control formulation is one, which is economical to produce, safe, stable in the environment, easily delivered and should enable the bio-control agent to act effectively and consistently under varied environmental conditions. The nutrient status of the substrate or additives is one of the important factors influencing the efficacy of the formulation. Introduction of antagonists through organic carriers alleviate the competition from autochthonous microorganisms, since the organic carriers serve as protectant and a food base during the establishment of the antagonist (Deacon, 1988, Steinmetz and Schonbeck, 1992). Commercial formulations of *Trichoderma* are being widely used for disease management in agricultural and horticultural crops (Table 2). Delivering of *Trichoderma* spp. to the site of infection via the organic food bases like diatomaceous earth granule with molasses (Backman and Rodriguez kabana, 1975), wheat bran (Henis *et al.*, 1978), wheat bran + saw dust (Elad *et al.*, 1980), wheat bran + peat (Sivan *et al.*, 1984), molasses yeast (Papavizas *et al.*, 1984), alginate pellets (Fravel *et al.*, 1985), pyrax (Papavizas and Lewis, 1989), vermiculite + wheat bran (Lewis *et al.*, 1991, Nakkeeran *et al.*, 1997) and processed manure pellets (Kok *et al.*, 1996) allow establishment and multiplication of the antagonist in the soil resulting in high population levels during an extended period of time.

Table 2: Commercial products from *Trichoderma* species for disease management

Product	Target pathogens	Crops recommended	Manufacturer
Bio-fungus- *Trichoderma* spp.	*Sclerotinia, Phytophthora, R. solani, Pythium* spp., *Fusarium, Verticillium*	Flowers, Strawberries, Trees, Vegetables	De Cuester, Belgium
Binab T- *T. harzianum* and *T. polysporum*	Pathogens causing wilt, take all, root rot, and internal decay of wood products	Flower, fruits, ornamentals, turf grass and vegetables	Bio-innovation Lab-UK
Root-Pro-*T. harzianum*	*R.solani, Pythium spp., Fusarium* spp., and *S.rolfsii*	Flowers and vegetables	Mycontrol Ltd., Israel
Root shield- *T. harzianum*	*R.solani, Pythium* spp., *Fusarium* spp	Trees, shrubs, & vegetables	Bioworks , Inc, USA
Supresivit- *T. harzianum*	Several fungal pathogens	Broad spectrum	Borregaard Bio Plant, Denmark
T22G, T22 planter box- *T. harzianum*	*R. solani, Pythium* spp., *Fusarium* spp., and *Sclerotinia homeocarpa*	Bean, cabbage, corn, cotton, cucumber, peanut, potato, sorghum, soybean, tomato and sugarbeet.	Bioworks, Inc, USA
Trieco- *T. viride*	*R. solani, Pythium* spp., *Fusarium* spp.,	Cotton, chillies, tomato, sugarcane, citrus, grapes, sunflower, cereals, and vegetables	Ecosense labs (1) Pvt. Ltd., Mumbai, India

Product	Target pathogens	Crops recommended	Manufacturer
Trichodex- *T. harzianum*	*B. cinerea, Colletotrichum* spp., *Fulvia fulva, Monilia laxa, Plasmopara viticola, Pseudoperonospora cubensis*	Cucumber, grapes, nectarine, soybean	Makhteshim Chemical Works Ltd., New York
Trichopel, Trichoject, Trichodowel, Trichoseal *T. harzianum* and *T. viride*	*Armillaria, Botryosphaeria, Chondrostereum, Fusarium, Nectria, Phytophthora, Pythium, Rhizoctonia*	Broad spectrum	Agrimm Technologies Ltd., New Zealand
Trichoderma 2000- *Trichoderma* spp.	*R. solani, Pythium* spp., *Fusarium* spp., and *S. rolfsii*	Nursery and field crops	Mycontrol Ltd., Israel

(Kannaiyan *et al.*, 2002)

Talc based formulation

Tamil Nadu Agricultural University, Coimbatore, has developed the technology of mass production of talc-based formulation of *T. viride* and *Pseudomonas fluorescens* for seed treatment and soil application. Several private industries in Coimbatore, Bangalore, Chethali, Delhi and Chennai produce the same in large quantities. However, the demand exceeds the present supply. The annual requirement of *Trichoderma* has been estimated as 5,000 tonnes to cover 50 per cent area in India. Indirectly it also creates self-employment opportunities to the unemployed youth. The first commercial seed treatment formulation of bio-control agent in India was developed by Jeyarajan *et al.* (1994). *T. viride* was grown in molasses yeast medium for 10 days in flasks. Then 2.5 l of the culture were inoculated to 50 l of sterilized molasses yeast medium in fermentor. It was incubated for 10 days with 4-8 h aeration/day. The fungal biomass and the broth were mixed with 100 kg talc (super white) powder and 500 g of carboxy methyl cellulose as a sticker. It was dried in shade for 72 h and packed in alkathene bags. The initial population of *Trichoderma* in the produce was 300 x 10^6 cfu/g. The product should contain a minimum of 20 x 10^6 cfu/g at the time of use. This is applied @ 4 g per kg of seed. The shelf life was 4 months. Angappan (1992) found that chickpea treated with this product maintained the rhizosphere population at 11 to 13 x 10^3 cfu/g of soil throughout crop growth.

Other formulations

Ranganathan *et al.* (1995) found that gypsum was a good and cheaper substitute for talc. Nakkeeran and Jeyarajan (1996) tested two industrial waste namely precipitated silica and calcium silicate in the place of talc and found them to give a population of 99, 104 x 10^6 cfu/g respectively compared to 143 x 10^6 cfu/g in talc formulation after 4 months of storage. These two substrates were also much cheaper than talc.

Diatomaceous earth granules were added into a broth consisting 100ml black grape molasses, 900 ml of water, 3 g each of KNO_3 and KH_2PO_4 till the level of saturation. It was autoclaved at 121°C for 15 min and spread in shallow pans to a height of 3-5 cm and autocalved again. *T. harzianum* (3 day old culture) was homogenized in a blender for 3 seconds and mixed with sterilized granules in shallow pans and incubated at 25°C for 4-7 days. Clumps were broken up and granules air dried with frequent stirring and used as an inoculum. The inoculum was mixed at 1:1 (v/v) with sterilized diatomaceous earth granules impregnated with 10 per cent molasses solution. It was applied to peanut at 140 kg/ha on 70 and 100 days after sowing to control *S. rolfsii*. The disease was reduced by 42 per cent over control and yield increased by 13.5 per cent (Backman and Rodriguez Kabana, 1975).

Wheat bran: Sawdust: tap water mixture (3:1:4 v/v) is taken in polypropylene bags and autoclaved for 1 h at 121°C for two successive days. The bags were inoculated with *T. harzianum* and incubated in illuminated chambers for 14 days at 30°C. It was applied at the time of sowing and mixed with the soil to a depth of 7-10 cm with a rotary hoe. It increased yield of beans (1500 kg/ha), tomato (300 kg/ha), cotton (500 kg/ha) and potato (40-600 kg/ha) and controlled *S. rolfsii* and *R. solani* (Elad *et al.*, 1986).

Wheat bran:Peat mixture (1:1 v/v) was autoclayed for 1h. Substrate moisture was adjusted to 50 per cent (w/w) with sterile water, medium was inoculated with 0.1 ml of a conidial suspension containing 2 x 10^4 conidia/ml and incubated for seven days at 30°C. It was mixed with rooting mixture for tomato at 10 per cent v/v to control crown rot (*Fusarium oxysporum f.sp. radicis lycopersici*). The disease was reduced by 29.7 per cent over control and yield increased by 7 per cent (Sivan *et al.*, 1984).

Trichoderma was multiplied in molasses—yeast medium for 10 days. Vermiculite (Grade 4) and milled wheat bran (250 mesh) were heated by placing over metal pans at 70 °C in hot air oven for three days. Vermiculite (100g), wheat bran (3.3 g), liquid culture (14 ml) and 0.05N HCl (17.5 ml) were mixed and packed. This was immediately used for soil application (Lewis *et al.*, 1991).

Sodium alginate (20g) was dissolved in 750 ml water at 40°C on a stirring hot plate. Wheat bran ground to pass through a 0.425 mm mesh screen was placed in a glass blender container with distilled water (50g/250 ml). Kaolin can also be used in place of bran. It was autoclaved for 30 min and cooled. Fermentor biomass (16-21 g/l) was added to provide 7 x 10^6 conidia and chlamydospores. The mixture containing fungus, alginate and bran or kaolin was added drop wise into 500 ml gellant solution (0.25 M $CaCl_3$, pH 5.4). As it entered, each droplet gelled and a distinct spherical bead formed. After 20 min in the gellant, beads were separated from the solution by gentle filtration, washed and dried for 24 h in a stream of air at 25°C (Lewis and Papavizas, 1986).

Shelf life

Shelf life of a bio-control agent plays a crucial role in adoption of the technology. In general the antagonists multiplied in a organic food base has greater shelf life than the inert or inorganic food bases. Viability of the encapsulated alginate pellets of *T. harzianum* isolate (Thr IDI) remained viable for at least 6 months when stored at 5°C and were reduced significantly when stored at 2°C (Dandurand and Knudsen, 1993). Jeyarajan *et al.* (1994) developed talc, peat, lignite and kaolin based formulation of *T. viride*, which had a shelf life of 4 months. Shelf life of gypsum-based formulation was also four months, but was cheaper than talc based formulations (Ranganathan *et al.*, 1995). Vermiculite bran acid fermentor biomass (VBA-FB) of *T. viride* recorded the highest mean population of 20.5 x 10^7 cfu/g up to 75 days of storage at room temperature (Nakkeeran *et al.*, 1997).

Delivery system

Biocontrol agents can be applied to various plant parts like seed, root, foliage, inflorescence and into the soil. Several effective methods of delivery system have been developed which enhanced the yield of crops considerably (Nakkeeran *et al.* 1995; Nakkeeran and Renukadevi, 1997).

Seed treatment

Seed treatment is one of the methods for introducing bio-control agents into soil-plant environment. The efficacy of seed treatment depends on the ability of the antagonist to multiply in the rhizosphere region. Seed treatment with *T. harzianum* reduced the disease caused by *R. solani* on cotton under field conditions (Elad *et al.*, 1982). Treatment of black gram, green gram, pigeon-pea, cowpea, groundnut, sunflower, sesame and cotton with *T. viride* @ 4g/kg reduced the incidence of root rot and wilt (Dinakaran *et al.*,1995; Jeyarajan *et al.*, 1994: Jeyarajan and Nakkeeran, 1996; Muthamilan and Jeyarajan, 1996; Nakkeeran and Renukadevi, 1997).

Solid matrix priming

It is a process in which moistened seeds are mixed with an organic carrier and the moisture content of the mixture is brought to a level just below that required for seed sprouting. Solid matrix priming can further enhance the efficacy of antagonist in treated seeds as it grows on seed surface during priming process and thus the propagule numbers increase. Slurry of *Trichoderma* is first added to seeds followed by addition of lignite and water. *Trichoderma* required less moisture for growth. The primed seeds were incubated for 4 days at 80-100 per cent RH. During this process *Trichoderma* profusely sporulate on seed surface. In *Pythium ultimum* infested soil the stand was increased to 70-80 per cent as against 10 per cent in control. In tomato seed the population of *Trichoderma* increased by ten fold due to solid matrix priming (Harman *et al.*, 1989).

Soil application

High population of the antagonist in the soil is necessary for effective management of soil borne pathogens. Adams (1990) defined efficiency of biocontrol agents as the ratio of number of propagules of mycoparasites required to obtain disease control to the typical inoculum density of a plant pathogen. Therefore, attempts were made to develop suitable delivery system to soil. For controlling *R. solani*, *Trichoderma* population of 5 x 10^6 was required for each propagule of *R. solani*. Addition of wheat bran based inoculum to soil gave 80 per cent reduction of root rot over control in chickpea and bean (Elad *et al.*, 1986). Incidence of urdbean root rot was reduced by 91.3 per cent by adding *T. viride* + *T. harzianum* to soil (Kousalya and Jeyarajan, 1988). Delivering *T. harzianum* through soil during sowing increased the survival of peanut (90 per cent), while in control none of the plants survived (Muthamilan and Jeyarajan, 1996).

Cut stump application

Ricard (1981) developed a special shears to apply *T. viride* simultaneously with pruning to protect fruit trees against silver leaf disease caused by *Chondrostereum purpureum*. The shears have a reservoir for suspension of *T. viride*. Each pruning cut was covered by spore suspension.

Hive insert

Thomson *et al.* (1992) developed an innovative method of application of bio-control agent right in the infection court at the exact time of susceptibility. *Botrytis cinerea* (fruit rot of straw berry) was effectively controlled by *T. virens* through honey bees which acted as vehicle to carry the spores of *Trichoderma*. Peng *et al.* (1992) developed a novel and more efficient method of applying *T. virens* on talc and corn meal (5 x 10^8 cfug/g), which was placed inside the beehive in a dispenser. Bees acquired the conidia on their legs and bodies as they crawled. Each bee had a spore load of 88 to 1800 x 10^3. Each flower got 1.6 to 27 x 10^3 conidia of antagonist. It suppressed *B. cinerea* on stamens (54 per cent) and petals (47 per cent) and controlled fruit rot.

Population dynamics and its importance

Soil

The soil ecosystem is a hostile environment for introduced microorganisms, which often cannot establish and eventually disappear. The bio-control efficacy of *Trichoderma* isolates seems to depend on the population size, reached after introduction into soil. A successful antagonist, to survive better in soil and rhizosphere, should have the ability to proliferate well in the introduced environment (Baker and Cook, 1974). Population densities of *T. viride* (T-1-R4) and *T. harzianum* (WT-6-24) increased to 10^4 and 10^3 fold respectively in natural soil during the first 3 weeks of incubation (Lewis and Papavizas, 1986). The decline in the population of *T. viride* in the soil might be due to lysis and disintegration of conidia. The stable population densities of *T. viride* from 9-36 weeks of incubation may be due to the survival of chlamydospores. In no case did conidia of *Trichoderma* and *Gliocladium* proliferated in soil without the addition of a food base (Papavizas, 1985). Pyrax (Clay powder), alginate pellets, vermiculite-bran mixture based formulations allowed establishment and multiplication of the antagonist in soil, resulting in high population levels during the extended period of time. The activity and proliferation of *Trichoderma* spp. in the protection of radish seeds against *R. solani* was effective only in acidic soils (Chet and Baker, 1980; 1981). A food base such as ground wheat bran enables the fungal propagule in a given preparation to germinate and to colonize the soil. In sterile conditions the propagation of *T. harzianum* was much greater than under un-sterile soils. The incorporation of *T. harzianum* conidia through processed manure pellets helped the colonization of non-sterile soils indicated the necessity of food base for establishment. The soil pH also influenced the population dynamics (Kok *et al.*, 1996).

Seeds

Bio-control agents must proliferate rapidly and become established on the place where they are applied. The first 12-24 h after planting is a crucial period for the bio-protectant applied as seed treatment against soil-borne pathogens and they need to be grown and established on the seed coat (Hubbard *et al.*, 1983; Taylor and Harman, 1990). Solid matrix priming is the process, in which seeds are moistened and mixed with a finely ground organic carrier, sufficient water was added to achieve the appropriate moisture potential for priming and then incubated for a given duration at a constant temperature. Organic carriers have been shown to enhance the efficacy of *Trichoderma* spp. proliferate during the incubation period and become established, more active or have colonized the spermosphere thoroughly before planting (Harman *et al.*, 1981). Primed seeds consistently emerge more rapidly than the untreated seeds and provide better stands under adverse environmental conditions than do untreated seeds (Harman and Taylor, 1988). Solid matrix priming of tomato seeds increased the number of *Trichoderma* propagules by one order of magnitude in the spermosphere. The effectiveness of bio-control agents may depend partially on their ability to proliferate during a short period of favorable environmental conditions before they encounter plant pathogens. The number of colony forming units (cfu) of *Trichoderma* increased on seeds during the matrix priming in Agro-lig by about 10 fold. Number of *Trichoderma* on matrix priming were 10^4 cfu/ seed after four days of incubation, while on seeds treated with *Trichoderma* alone were about 10^3 cfu/seed. Hence the increase in the cfu of *Trichoderma* provided better emergence of tomato (Harman and Taylor, 1988).

Industrial application of *Trichoderma*

Trichoderma, a wood degrading fungus of the forest has well defined extracellular enzyme production and it is widely used in textile and food industries. It is used commercially for the production of cellulases and other enzymes that degrade complex polysaccharides. The cellulases are also used in "biostoning" of denim fabrics to give rise to the soft, whitened fabric stone-washed denim. In addition the enzymes are used in poultry feed to increase the digestibility of hemicelluloses from barley or other crops. Apart from it, *Trichoderma* is also used in industries to increase ethanol production (Stevenson and Weimer, 2002). Hence, the concept of transforming renewable plant biomass to fuels attracts attention. In semi solid fermentation, use of cellulose and 8% cellulase (from *T. reesei*) increased ethanol production.

Trichoderma, a soil health promoter

Lignocelluloses being the major source of carbon to ecosystems and the principal constituent of agricultural, forestry and municipal wastes, cellobiohydrolases and endoglucanases produced by *Trichoderma* spp., act on the lignocellulose and aids in the recycling nutrients by decaying them. It not only cleans the environment, but also improves the mineral pool in the soil and thereby increase the soil health and plant health. *T. harzianum* 1295-22 solubilized manganese dioxide, Zn metal and rock phosphate by chelation and reduction. In addition, it also aids in recycling carbon and other elements by decaying organic matter and dead trees in forests (Altomare *et al.*, 1999).

Bioremediation by *Trichoderma*

The variety of organic pollutants released by the ever increasing number of industries is the direct cause of environmental and health-related problems that have detrimental effects on living beings. Conventional approaches (*e.g.* land-filling, recycling, pyrolysis and incineration) to the remediation of contaminated sites are inefficient and costly and can also lead to the formation of toxic intermediates. Thus, biological decontamination methods are preferable to conventional approaches because, in general, microorganisms degrade numerous environmental pollutants without producing toxic intermediates.

Different organic pollutants that are present in the environment persist for varying periods of time. Depending on their half-life, some pollutants either disappear within a short period of time or transformed into non-toxic end products. Many others, however, such as polychlorinated dibenzodioxyfurans (PCDDF/Fs), dioxins, dichlorodiphenyltrichloroethane (DDT), chlordane and hexachlorocyclohexane (HCH) have long half lives in the environment and tend to enter the food web, where they are subsequently 'biomagnified'. On entering mammalian systems, most of these pollutants are often metabolized to generate more toxic and harmful intermediates owing to the nonspecific activity of different mammalian enzymes. Bioremediation is the process by which living organisms degrade or transform hazardous organic contaminants. *Trichoderma* plays a vital role in bioremediation of heavy metals, toxic chemicals such as xenobiotic compounds, cobalt, lignite, cyanide, nitrocellulose etc.

Trichoderma spp. are recognized as "keystone or controller organisms" in forest soil due to their involvement in decomposition, nutrient cycling, and their regulation of associated mycobiota. Recent evidences have proved their potential to decompose xenobiotics. *T. harzianum* and *T. viride* have been shown to be capable of degrading organochlorides under *in vitro*. The predominant association of *T. harzianum* and *T. viride* may influence the persistence of organochloride contaminants in forest soil (Smith, 1995). *T. harzianum* was capable of degrading chlorophenols and adsorbable organic halogens. It also had the capability to degrade ciliatine, glyphosate, and also phosphonic acid (Lupicka *et al.*, 1997).

Biodegradation of noxious chemicals

Trichoderma species also have the potential to degrade noxious chemicals like nitrocellulose, cobalt 60, cyanides and heavy metals. *T. reesei* had the capacity to degrade nitrocellulose which is commonly used in explosives and pharmaceutical industries. The fungus was able to degrade nitrocellulose completely after 22 days of application (Auer *et al.*, 2005). Cotton, beans or corn-grown on a polluted site (*i.e.* heavy metals, arsenic, cyanide and metallocyanides, nitrates, etc.) in association with *T. harzianum* increased the uptake of nutrients or toxicants by the plants due to the production of enzymes like rhodanase, formamide hydro lyase and cyanide hydratase (Ezzi and Lynch, 2005).

Biodegradation of pesticide

T. harzianum degraded DDT, dieldrin, endosulfan, PCNB, and pentachlorophenol. *T. harzianum* CCT-4790 degraded 60 per cent of the herbicide diuron in soil and opens a good potential for bioremediation of herbicide-contaminated soils. *T. viride* degraded pesticide photodieldrin into water soluble and non-insecticidal compounds after 4 to 5 weeks. Thus, *Trichoderma* strains play a crucial role in bioremediation process. *T. harzianum*, a biocontrol species colonize plant rhizosphere and it metabolize several pesticides. Pentachlorophenol (PCP) was metabolized to pentachloroanisole (Rigout and Matsumura, 2002). *Trichoderma* is a predominant microflora in the rhizosphere of oil polluted soils of Kuwait. It encouraged remediation of oil in the rhizosphere region of polluted soil (Lynch and Moffat, 2005).

Phytobiation

Physical and chemical techniques are the most commonly used methods for remediation of pollutants. The high costs of these methods have stimulated interest in alternative biologically-based systems, in particular, bioremediation and phytoremediation. Bioremediation consists of adding microorganisms to a polluted system, in the presence of nutrients, for degrading toxic compounds. Phytoremediation uses plants to degrade or remove toxicants from soil lacking a defined or managed root-microbial population. Combined application of both remediation is termed as phytobiation. Cultivation of cotton, beans or corn as well as ferns on a polluted site rich in heavy metals, arsenic, cyanide and metallocyanides, nitrates, etc. in association with *Trichoderma* spp. increased rapid uptake of the toxic elements by the plants. In addition there was strong reduction of toxicity caused by polycyclic aromatic hydrocarbons and phenolic compounds in water and soil by the use of *Trichoderma* (Vinale *et al.*, 2003).

Future thrust

The investigation of the biogeographical distribution of fungi is a challenging task. Firstly, it is generally regarded that only a small proportion of fungi (approximately 1,00,000 species) have been identified to date, with estimates of the numbers of fungi at 1.5 million. This estimate is not universally accepted, as it is based on calculations from temperate regions indicating an approximately 6: 1 ratio of fungal species to plant species. However, for tropical regions there are conflicting estimates, with this ratio either considered too high or too low or proportions of up to 33: 1 considered more realistic. Consequently, the incomplete knowledge of the diversity and distribution of fungi has to be taken into account in any fungal biogeographic study. Owing to the significant contribution of *Trichoderma* to human welfare and due to its rich biodiversity in India, the following issues may be addressed to conserve and to claim the IPR rights of our microbial wealth. They are:

- Adoption of polyphasic systematic to map the distribution of *Trichoderma* from different Agro-climatic zones.
- Identifying the habitat preference of individual taxa.
- Exploring the potential of all the strains of *Trichoderma* for human welfare.
- Research on bioremediation, phytoremediation, phytobiation has to be strengthened to select potential strains of *Trichoderma*.
- Transformation and gene manipulation strategies have to be strengthened for exploring the potential of *Trichoderma* other than *T. reesei*.
- Strengthening proteomic and genomic research of *Trichoderma* for better pest and disease control of plants.
- Improving the knowledge on the ecology of *Trichoderma*—plant-pathogen interaction.
- Assessment of human and plant health risks.

- Human Resource Development on bio-informatics, systems biology, digital biology with molecular techniques for the documentation and exploitation of the full potential of *Trichoderma* for the welfare of the country and the mankind as a whole.

References

Adams, P.B. 1990. The potential of microparasites for biological control of plant diseases. *Ann. Rev. Phytopathol.*, **28**: 59-72.

Altomare, C., Norvell, W.A., Bjorkmann, T. and Harman, G.E. 1999. Solubilization of phosphates and micronutrients by the Plant Growth Promoting and Biocontrol fungus, *Trichoderma harzianum* Rifai 1295-22. *Appl. Environ. Microbiol.*, **65**(7): 2926-2933.

Angappan, K. 1992. *Biological control of chickpea root rot caused by Macrophomina phaseolina (Tassi.) Goid.* M.Sc.(Ag) thesis, TNAU, Coimbatore, p. 114.

Auer, N., Hedger, J.N. and Evans, C.S. 2005. Degradation of nitrocellulose by fungi. *Biodegradation*, **16**(3): 229-236.

Backman, P.A. and Rodrigue-Kabana, R. 1975. A system for the growth and delivery of biological control agents to the soil. *Phytopathology*, **65**: 819-821.

Baker, K.F. and Cook, R.J. 1974. *Biological control of plant pathogens.* W.H. Freeman and Co., San Francisco, C.A. p. 433.

Bissett, J. 1991a. A revision of the genus *Trichoderma*. II. Infrageneric classification. *Can. J. Bot.*, **69**: 2357-2372.

Bissett, J. 1991b. A revision of the genus *Trichoderma*. III. Sect. *Pachybasium*. *Can. J. Bot.*, **69**: 2373-2417.

Bissett, J. 1991c. A revision of the genus *Trichoderma*. IV. Additional notes on section *Longibrachiatum*. *Can. J. Bot.*, **69**: 2418-2420.

Chet, I. and Baker, R. 1980. Induction of suppressiveness to *Rhizoctonia solani* in soil. *Phytopathology*, **70**: 994-998.

Chet, I. and Baker, R.1981. Isolation and biocontrol potential of *Trichoderma hamatum* from soil naturally suppressive to *Rhizoctonia solani*. *Phytopathology*, **71**: 286-290.

Connick, Jr., W.J., Lewis, J.A. and Quimby Jr., P.C. 1990. *Formulation of biocontrol agents for use in Plant Pathology, New Directions in Biological control. Alternatives for Suppressing Agricultural Pests and Diseases* (Eds. R.R. Baker and P.E. Dunn). Alan R. Liss, Inc. New York, pp. 345-372.

Dandurand, L.M., and Knudsen, G.R. 1993. Influence of *Pseudomonas fluorescens* on hyphal growth and biocontrol efficacy of *Trichoderma harzianum* in the spermosphere and rhizosphere of pea. *Phytopathology*, **83**: 265-270.

Danielson, R.M and Davey, C.B. 1973. The abundance of *Trichoderma* propagules and the distribution of species in forest soils. *Soil Biol. Biochem.*, **5**: 486-494.

Deacon, J.W. 1988. Biocontrol of soil-borne plant pathogens with introduced inocula. *Phil. Trans. R. Soc. Lond. Ser. B.*, **318**: 249-264.

Dennis, C. and Webster, J. 1971. Antagonistic properties of species groups of *Trichoderma* I. Production of non-volatile antibiotics: *Trans. Br. mycol., Soc.* **57**: 25-39.

Dhingra, O.D. and Khare, M.N. 1973. Biological control of *Rhizoctonia bataticola* on Urdbean. *Phytopath. Z.*, **76**: 23-29.

Dinakaran, D. and Marimuthu, T. 2001. Effect of *Trichoderma viride* mutants on the management of root rot of blackgram. *Natl. Symp. on Pulses and Oilseeds for Sustainable Agriculture,* July 29-31, 2001, Directorate of Research, Tamil Nadu Agricultural University, Coimbatore, India, pp. 181. (Abst.).

Dinakaran, D., Ramakrishnan, G., Sridar, R. and Jeyarajan, R. 1995. Management of sesamum root rot with biocontrol agents. *J. oilseeds Res.*, **12**: 262-263.

Elad, Y., Chet, I. and Henis, Y. 1982. Degradation of plant pathogenic fungi by *Trichoderma harzianum. Can. J. Microbiol.,* **28**: 719-725.

Elad, Y., Chet, I. and Katan, P. 1980. *Trichoderma harzianum.* A biocontrol agent effective against *Sclerotium rolfsii* and *Rhizoctonia solani. Phytopathology,* **70**: 119-121.

Elad, Y., Zuiel, Y. and Chet, I. 1986. Biological control of *Macrophomina phaseolina* (Tassi) Goid. by *Trichoderma harzianum, Crop Prot.*, **5**: 288-292.

Elisa Esposito and Manuela da Silva.1998. Systematics and environmental application of the genus *Trichoderma. Critical Rev. Microbiology.*, **24**: 89-98.

Ezzi, M.I. and Lynch, J.M. 2005. Cyanide catabolizing enzymes in *Trichoderma* spp. *Journal of Enzymes and Microbial Technology,* **31**: 1042-1047.

Fravel, D.R., Marois, J.J., Lumsden, R.D. and Connick Jr., W.J. 1985. Encapsulation of potential biocontrol agents in an alginate – clay matrix. *Phytopathology,* **75**: 774-777.

Grondona, I., Hermosa, R., Tejada, M., Gomis, M.D., Mateos, P.F., Bridge, P.D., Monte, E. and Garcia-Acha, I. 1997. Physiological and biochemical characterization of *Trichoderma harzianum*, a biological control agent against soilborne fungal plant pathogens. *Appl. Environ. Microbiol.*, **63**: 3189.

Harman, G.E. and Taylor, A.G. 1988. Improved seedling performance by integration of biological control agents at favourable pH levels with solid martix priming. *Phytopathology,* **78**: 520-525.

Harman, G.E., Chet, I. and Baker, R. 1981. Factors affecting *Trichoderma hamatum* applied to seed as a biocontrol agent. *Phytopathology,* **71**: 569-572.

Harman, G.E., Taylor, A.G. and Stasz, T.E. 1989. Combining effective strains of *Trichoderma harzianum* and solid martix priming to improve biological seed treatment. *Plant Dis.,* **73**: 631-637.

Henis, Y., Ghaffar, A. and Baker, R. 1978. Integrated control of *Rhizoctonia solani* damping-off of radish. Effect of successive plantings. PCNB and *Trichoderma harzianum,* pathogen and disease. *Phytopathology,* **68**: 900-907.

Hjeljord, L. and Tronsmo, A. 1998. *Trichoderma* and *Gliocladiumin* Biological Control: An Overview. In*: Harman, G.E., Kubicek, C.P. (Eds.), Trichoderma and Gliocladium. Vol. 2. Enzymes, Biological Control and Commercial Applications.* Taylor and Francis Ltd., London, pp.131-151.

Hubbard, J.P., Harman, G.E. and Hadar, Y. 1983. Effect of soil-borne *Pseudomonas* spp. on the biological control agent, *Trichoderma hamatum*, on pea seeds. *Phytopathology*, **73**: 655-659.

Jeyarajan, R. and Nakkeeran, S. 1996. *Exploitation of biocontrol potential of Trichoderma for field use In: Current trend in life sciences vol. XXI* (eds. K.Manibhushan Rao and A. Mahadevan) Today and Tomorrow's Printers and Publishers, New Delhi, India. pp. 61-66.

Jeyarajan, R.G., Ramakrishnan, Dinakaran, D. and Sridar, R. 1994. Development of products of *Trichoderma viride* and *Bacillus subtilis* for biocontrol of root rot diseases. In: *Biotechnology in India* (ed. Dwivedi) Bioved Research Society, Allahabad, pp. 25-36.

Kannaiyan, S., Sabitha Doraiswamy, Ramakrishnan, G., Rabindra, R.J., Samiyappan, R., Udayasurian, V., Gunasekaran, K., Kennedy, J.S., Nakkeeran, S. and Sathiah, N. 2002. *Recent Advances in Eco-friendly Approaches for the Management of Pests and Diseases in Crops.* Tamil Nadu Agricultural University, Coimbatore. pp. 1-389.

Kok, C.J., Hageman, P.E.J., Maas, P.W.T., Postma, J., Roozen, N.J.M. and Van Vuurde, J.W.L. 1996. Processed manure as carrier to introduce *Trichoderma harzianum* Population dynamics and biocontrol effect on *Rhizoctonia solani. Biocontrol Sci. and Tech.*, **6**: 147-161.

Kousalya, G. and Jeyarajan, R. 1988. Techniques for mass multiplication of *Trichoderma viride Pers. Fr* and *T. harzianum Rifai. National Seminar on management of crop diseases with plant products/biological agents. 10-12 January,* Agricultural College and Research Institute, Madurai. pp. 32-33.

Kredics, L., Antal, Z., Doczi, I., Manczinger, L., Kevei, F. and Nagy, E. 2003. Clinical importance of the genus *Trichoderma*. A review. *Acta Microbiol. Immunol. Hung.*, **50**: 105-17.

Krishnamoorthy, A.S. 1987. *Biological control of damping of off disease of tomato caused by Pythium indicum* M.Sc(Ag) Thesis, Tamil Nadu Agricultural University, Coimbatore. p. 125.

Kubicek, C.P., Bissett, J., Druzhinina, I., Kullnig-Gradinger, C.M. and Szakacs, G. 2003. Genetic and metabolic diversity of *Trichoderma*: a case study on South East Asian isolates. *Fungal Genet. Biol.*, **38**: 310-319.

Kubicek, C. P. and Penttila, M.E. 1998. Regulation of Production of Plant Polysaccharide Degrading Enzymes by *Trichoderma*. In: Harman, G.E., Kubicek, C.P. (Eds.), *Trichodermaand Gliocladium.* Vol. 2. Enzymes, Biological Control and Commercial Applications. Taylor and Francis Ltd., London, pp. 49-71.

Kullinig, C., Szakacs, G. and Kubicek, C.P. 2000. Molecular identification of *Trichoderma* from Russia, Siberiya and the Himalaya. *Mycol. Res.*, **104**(9): 1117-1125

Kullnig-Gradinger, C.M., Szakacs, G. and Kubicek, C.P. 2002. Phylogeny and evolution of the fungal genus *Trichoderma:* a multigene approach. *Mycol. Res.*, **106**: 757-767.

Kumar, A. and Marimuthu, T. 1993. Biological control of damping-off of *Eucalyptus camaldulensis* caused by *Rhizoctonia solani* with *Trichoderma viride* and decomposed coconut coir pith. *Pl. Dis. Res.*, **9**(2): 116-121.

Lewis, J.A and Papavizas, G.C. 1986. Reduced incidence of *Rhizoctonia* damping off of cotton seedlings with preparations of biocontrol fungi. *Biol. Cult.Tests*, **1**: 48.

Lewis, J.A., Papavizas, G.C. and Lumsden, R.D. 1991. A new formulation system for the application of biocontrol fungi to soil. *Biocontrol Sci. Technol.,* **1**: 59-69.

Liu, S. and Baker, R. 1980. Mechanism of biological control in soil suppressive to *Rhizoctonia solani. Phytopathology*, **70** : 404-412.

Lupicka, K., Strof, W., Kubs, K., Skorupa, M., Wieczorek, P., Lejczak, B. and Kafarski, P. 1997. The ability of soil –borne fungi to degrade organophosphanate carbon to phosphorus bounds. *Appl. Microbiol. Biotechnol.*, **48**: 549.

Lynch, J.M. and Moffat, A.J. 2005. Bioremediation-prospects for the future application of innovative applied biological research. *Annals Appd. Biol.*, **146**(2): 217-221.

Marshall, D.S. 1982. Effect of *Trichoderma harzianum* seed treatment and *Rhizoctonia solani* inoculum concentration on damping off of snapbean in acidic soil. *Plant Dis.*, **66**: 788-789.

Muthamilan, M.R. and Jeyarajan, R. 1996. Integrated management of *Sclerotium* root rot of groundnut involving *Trichoderma harzianum Rhizobisum* and *Carbendazim. Indian J. Mycol. Pl. Pathol.,* **26**: 204-209.

Nakkeeran, S. and Jeyarajan, R. 1996. Exploitation of antagonistic potential of Trichoderma for field use. Paper presented in National Symposium on disease of plantation crops and their management. Institute of Agriculture, Sriniketan held during February, pp. 18-20, 1996.

Nakkeeran, S. and Renukadevi, P. 1997. Seed borne microflora of pigeonpea and their management. *Pl. Dis.Res.,* **12**: 103-107.

Nakkeeran, S. and Sabitha Doraisamy. 2001. Genetic improvement of *Trichoderma viride* and assessment of its efficacy against soil borne pathogens. In: National Seminar on *Emerging Trends in Pests and Disease Management.* Center for Plant Protection Studies, TNAU, Coimbatore. p. 130.

Nakkeeran, S., Kousalya Gangadharan and Renukadevi, P. 1995. Efficacy of organic amendments and antagonists on root rot and wilt of pigeonpea. National Symposium on *"Organic Farming",* TNAU, Madurai. Oct. 27-28, pp. 141-142.

Nakkeeran, S., Sankar, P. and Jeyarajan, R. 1997. Standardization of storage conditions to increase shelf life of *Trichoderma* formulations. *J. Mycol. Pl. Pathol.,* **27**: 60-63.

Nelson, E.E. 1982. Occurrence of *Trichoderma* in a Douglas-fir soil. *Mycologia,* **74**: 280-284.

Papavizas, G. C. 1985. *Trichoderma* and *Gliocladium*: Biology, ecology and potential for biocontrol. *Ann. Rev. Phytopathol.,* **23**: 23-54.

Papavizas, G.C. and Lewis, J.A. 1989. Effect of *Trichoderma* and *Gliocladium* on damping- off and blight of snapbean caused by *Sclerotium rolfsii. Plant Pathol.,* **38**: 277- 286.

Papavizas, G.C., Dunn, M.T., Lewis, J.A. and Beagle-Ristaino, J. 1984. Liquid fermentation technology for experimental production of biocontrol fungi. *Phytopathology,* **74**: 1171-1175.

Peng, G., Sutton, J.C. and Kevan, P.G. 1992. Effectiveness of honey bees for applying the biocontrol agents *Gliocladium rosdeum* to strawberry flowers to suppress *Botrytis cinerea Can. J. Pathol.,* **14**: 117-129.

Ranganathan, K., Sridar, R. and Jeyarajan, R. 1995. Evaluation of gypsum as a carrier in the formulation of *Trichoderma viride. J. Biol. Control.,* **9**: 61-62.

Ricard, J.L. 1981. Commercialization of *Trichoderma* based myco-fungicide. Some problems and solutions. *Biocontrol News and Information,* **2**: 95-98.

Rigout, J. and Matsumura, F. 2002. Assessment of the rhizosphere competency and pentachlorophenol-metabolizing activity of a pesticide-degrading strain of *Trichoderma* harzianum introduced into the root zone of corn seedlings. *J. Env.Sci. and Health,* **37**: 201-210.

Rossman, A.Y. 1996. Morphological and molecular perspectives on systematics of Hypocreales. *Mycologia,* **88**: 1-19.

Samuels, G.J., Petrini, O., Kuhls, K., Lieckfeldt, E. and Kubicek, C.P. 1998. The *Hypocrea schweinitzii* complex and *Trichoderma* sect. *Longibrachiatum. Stud. Mycol.,* **41**: 1-54.

Sivan, A., Elad, Y. and Chet, I. 1984. Biological control. Effect of new isolates of *Trichoderma harzianum* on *Phythium aphanidermatum. Phytopathology,* **74**: 498-501.

Sivasithamparam, K., Ghisalberti, E.L. 1998. Secondary Metabolism in *Trichoderma* and *Gliocladium.* In: Kubicek, C.P., Harman, G.E. (Eds.), *Trichoderma and Gliocladium.* Vol. 1. Basic Biology, Taxonomy and Genetics. Taylor and Francis Ltd., London, pp. 139-191.

Smith, W.H. 1995. Forest occurrence of *Trichoderma* species: emphasis on potential organochloride (xenobiotic) degradation. *Ecotoxicol. Environ. Saf.,* **32**: 179.

Steinmetz, J. and Schonbeck, F. 1992. Applicability of different formulations of fungal antagonists for the control of soil-borne diseases. *Bulletin OILB-SROP*, **15**: 206-208.

Stevenson, D.M. and Weimer, P.J. 2002. Isolation and characterization of a *Trichoderma* strain capable of fermenting cellulose to ethanol, *Appl. Microbiol. Biotechnol.*, **59**: 721-726.

Taylor, A.G. and Harman, G.E. 1990. Concepts and technologies of selected seed treatments. *Ann. Rev. Phytopathol.*, **28**: 321-329.

Thomson, S.V., Hansen, D.R., Flint, K. and Vandenberg, S.D. 1992. Dissemination of bacteria antagonistic to *Erwinia amylovora* by honey bees. *Plant Disease*, **76**: 1052-1056.

Thrane, U., Poulsen, S.B., Nirenberg, H.I. and Lieckfeldt, E. 2001. Identification of *Trichoderma* strains by image analysis of HPLC chromatograms. *FEMS Microbiol Lett.*, **203**: 249-255.

Ushamalini, C., Nakkeeran, S. and Marimuthu, T. 2006. Development of Biomanure for the management of Turmeric rhizome rot. *Archives of Phytopathology and Plant Protection*, (Online) pp. 1-12.

Vinale, F., Abadi, K., Ruocco, M., Marra, R., Scala, F., Zoina, A., Woo, S. and Lorito, M. 2003. Remediation of pollution by using biological systems based on beneficial plant-microorgan-ISMS interactions. *J. Plant Pathol.*, **85**: 301.

Whipps, J. M. 1997. Developments in the biological control of soil-borne plant pathogens. *Botanical Research*, **26**: 1-134.

Xu, T. 1996. Advances of molecular biology of *Trichoderma* spp. *Zhenjun Xuebao*, **15**: 143.

Zhang, C.L., Druzhinina, I.S., Kubicek, P.C. and Xu, T. 2005. *Trichoderma* biodiversity in China: Evidence for a North to South distribution of species in East Asia. *FEMS Microbiology Letters*, **251**: 251-257.

Soil Microbiology and Soil Biotechnology : Vol. 1
Editor : S. Kannaiyan
Associated Publishing Co., New Delhi, 2010, pp. 188-196

13

Microorganisms and Environment

S. Kannaiyan[*]

Introduction

The concern about environmental safety and sustainability of land productivity is increasing among scientists and environmentalists. With the increasing popultions, it is also becoming clear that food security to the teeming millions will not be possible unless the available resources are efficiently utilized for increasing the productivity. The need for more intensive and economic agriculture production led to wide use of high doses of concentrated chemical fertilizers but insufficient use of organics led to negative results of decrease in fertility, soil structure and soil microorganisms. Soil organic matter is one of the important components of the soil and the dead plant and animal remains and the dead tissues of microorganisms form the main source of soil organic matter.

In healthy soil, the biotic and abiotic components covering organic matter including soil microorganisms and other soil life, mineral particles, soil air and water exists in a stage of dynamic equilibrium and regulate the ecosystem processes in mutual harmony by complementing and supplementing each other. When the soil is in good health, the population of soil fauna and flora multiplies rapidly which in turn will sustain the biochemical process of dissolution and synthesis at a high rate (Palaniappan and Annadurai, 1999).

The use of chemical fertilizers at high level for intensive agriculture system for increasing the crop yield per unit area is not only deleterious to the environment by affecting the soil system but injurious to human health. Under these circumstances biological fertilizers are renewable, cost effective, sustainable and no deleterious effects on soil health and environmentally safe (Kannaiyan, 2002).

Microorganisms have great potential of degrading pesticides used in agricultural practices and they are the potential biological system in maintaining long term soil fertility and soil health management (Kannaiyan, 2002). Microbial systems are dynamic biological agents plays significant role in decomposing the organic materials on earth and also widely exploited "Clean up" the environmental problems in nature.

* E-mail: skannaiyan @ hotmail.com
National Biodiversity Authority, 475, 9th South Cross Street, Kapaleeswarar Nagar, Neelankarai, Chennai - 600 041, Tamil Nadu.

The ability of microorganisms to uptake and accumulate heavy metals such as, Co, Cd, Zn, Mn, Cu, Pb, Ni, Hg, Ag, etc. is very well known. *Pseudomonas, Alcaligenes Aspergillus, Rhizopus Trichoderma, Saccharomyces, Rhodotorula, Aureobasidium, Ophiostoma, Spirulina, Oscillatoria* and *Plectonema* and *Thalassiosira pseudonema* have such ability to take up and accumulate or degrade the metals and some of them utilize for their metabolic activity and they have the biotechnological potential as biological agents of effluent detoxification (Kuiper *et al.*, 2004).

Bioremediation

Worldwide, contamination of soil and ground water is a serious environmental problem. The negative effects of pollutants on the environment and on human health are diverse and depend on the nature of the pollution. The search for alternative methods for excavation and incineration to clean polluted sites resulted in the application of bioremediation techniques (Kuiper *et al.*, 2004). Industrial growth and increased agricultural and domestic activities have led to an increase in the biogeochemical cycling of many elements, including heavy metals. This had resulted in significant amounts of heavy metals being deposited into natural aquatic and terrestrial ecosystems (Babich and Stotzky, 1978). Toxicity and in some cases commercial value, warrant metal recovery from aqueous solutions (Volesky, 1987).

The disposal, removal and treatment of various pollutants of air, water and land may not be a viable technology due to heavy cost involved in these process and other limitations. Also, some of the environmental pollutants that are to undergo biodegradation which leads to the biomagnification. Particularly, when spills occurs in sea waters and also when underground storage tanks of petrol begin to discharge hydrocarbons to subsoil and groundwater due to leakage, the situation of pollution becomes serious to use the water for drinking and also becomes unfit for irrigation to agricultural lands and affects the environment. It has been recognized quite recently that biological systems primarily the microbial origin could effectively prove potential means of degrading such complex environmental pollutants as well as preventing pollution through waste treatment both solid as well as liquid. Such recent environmental applications of biotechnology emerged into a new area of research and development called bioremediation.

Due *et al.* (2002) defined bioremediation as the action of microbes or other biological systems to degrade environmental pollutants. Bioremediation can be applied *in situ* without the removal and transport of polluted soil and without the disturbance of the soil matrix. Another advantage is that bacterial degradation usually results in complete mineralization of pollutant.

Important parameters for bioremediation are

- The nature of pollutants.
- The soil structure and hydrogeology—the movement of pollutants through soil and groundwater.
- The nutritional state and microbial composition of the soil (Due *et al.*, 2002).
- In many cases, the most effective solution combines several treatment techniques.

Bioremediation is the process of initiation of microorganisms to remove pollutants from the environment. It is an acceleration of the natural fate of biodegradable pollutants and hence a natural or green solution to the problem of oil pollutants. Bioremediation is defined as "the use of living organisms primarily the microorganisms to degrade environmental pollutants or to prevent pollution through waste treatment". Bioremediation is emerging as an ideal alternative technology for removing pollutants from the environment restoring contaminated sites and preventing further pollution. This environment friendly technology is expanding a range of organisms to be used to clean up pollution and forms a vital component of the so called "green movement" of maintaining the natural ecological balance which is an issue at present being top priority of environmental awareness and public policy (Kuiper *et al.,* 2004).

Bioremediation is a necessary and cost effective means of removing certain environmental pollutants that adversely affect human health or environmental quality. The enormous natural capacity of diverse microorganisms to degrade numerous organic compounds ranging from petroleum hydrocarbons to chlorinated solvents and to transform various inorganic compounds including metals, forms the basis for bioremediation. The metabolic activities of microorganisms are used to change an undesirable chemical into one that has less objectionable properties for example changing to toxic pesticide or a carcinogenic petroleum hydrocarbon into non-toxic carbon dioxide and water. At present most bioremediation projects have depended on naturally occurring microorganisms or native microflora from polluted sites, often the indigenous or native microflora in a contaminated and polluted sites. Currently many research groups are targeting for the selection of range of microoganisms available in different ecosystems including the use of genetically engineered microbes and also search for greater microbial diversity that can grow and adapt adverse conditions with innate capacity of pollutant degradation kinetics. It has been shown in Japan that a strain of *Pseudomonas* that can grow in solvents containing 50% toluene. Genetic engineering of such solvent resistant microbial strain to include various catabolic genes may result in the creation of a microbe that could degrade compounds in nonaqueous phase liquids that frequently contaminate soils (Kuiper *et al.,* 2004).

Pollutant-degrading microbial consortia

Consortium of bacteria each with different parts of the catabolic degradation route involved in the degradation of a certain pollutant often is found to be more efficient than the introduction of one single strain with the complete pathway. During rhizoremediation, the degradation of a pollutant in many cases is also the result of the action of consortium. The co-inoculation of *P. putida* PCL 1444 and *P. putida* PCL 1445 could be a significant role in the degradation of naphthalene by the conversion of 1, 2 dihydroxy-naphthalene into cis-o-hydroxy benzylpyruvic acid. The co-inoculation have shown that communities are formed on the roots in the presence of naphthalene presumably because *P. putida* PCL 1444 secretes naphthalene intermediates that are used by *P. putida* PCL 1445 when the strains are near each other on the root (Kuiper *et al.,* 2004).

Bioremediation is recognized as a suitable tool to restore contaminated sites. The combination of bioaugmentation and phytoremediation resulting in rhizoremediation could solve some of the problems encountered during the application of both individual techniques. It has been demonstrated that rhizoremediation can contribute to the restoration of polluted sites. It is now clear that suitable microbe-plant combinations can be selected for other pollutants. It may be concluded that the exploitation of suitable rhizosphere bacteria or fungi or actinomycetes or communities of organism would be able sustain and proliferate on the root system of a plant which is suitable for rhizoremediation or phytoremediation can yield useful novel-engineered systems. These systems then can be an interesting tool to further improve and develop bioremediation into widely accepted technique (Kuiper *et al.,* 2004).

Bioaugmentation

Bioaugmentation is a method to improve degradation and enhance the transformation rate of xenobiotics by the injection *i.e.* seeding of specific microbes able to degrade the xenobiotics of interest. Many microbes are described to have the genetic tools to mineralize recalcitrant pollutants such as PAH's (Polycyclic Aromatic Hydro carbons), chlorinated aliphatics and aromatics, nitroaromatics and long chain alkanes (Heidelberg *et al.,* 2002).

These microbes can be wild-type isolates but also can be genetically modified strains equipped with catabolic plasmids, containing the relevant degradation genes (Dixon, 1996). The analysis of the efficiency of microbial degradation showed that chlorinated derivatives are generally difficult to metabolize and frequently are degraded by means of co-metabolism (Dean-Ross, 1987). Co-metabolism is defined as the oxidation of non-growth substrates during the growth of an organism on another carbon or energy source (Kuiper *et al.,* 2004).

Construction of genetically modified strains is especially relevant in the case of pollutants such as polychlorinated biphenyls (PCBs) and chloroethylenes for which as far as is known, no microbial catabolic pathways have been evolved (Megharaj *et al.,* 1997) or no optimal pathways are available (Brazil *et al.,* 1995). Both the inability to improve the action microbial consortia and the restriction of degrading only a few pollutants limit the use of GEMs. In addition, legislative problems arise when GEMs are introduced in the environment (Sayler and Ripp, 2000). One example of a field release is the genetically engineered *Pseudomonas fluorescens* HK-44 (King *et al.,* 1990) which degrades naphthalene and was used to monitor the presence of naphthalene on a PAH-polluted site (Ford *et al.,* 1999). Sometimes the introduced microorganisms fails to degrade the soil and water pollutants due to various reasons including poor establishment of the inoculated microorganisms in the contaminated sites.

Goldsteing *et al.* (1985) reported five possible reasons for the failure of bacterial inocula:

- The concentration of the contaminants at a site can be too low to support growth of the inoculum. This also includes the problem of low bioavailability of certain pollutants.
- The presence of certain compounds in the environment can inhibit the growth or activity of inoculum.
- The protozoan-grazing rates on the inoculum can be higher than the growth rates of bacteria resulting in the decline of the bacterial popultion.

- The inoculum can prefer to use other carbon sources present in soil instead of contaminant.
- The inoculum may fail to establish because of the inability of the microbes to spread through the soil and reach the pollutant.

The inoculum size and the salinity level of soil also plays an important role in bioremediation. The importance of various parameters such as oxygen, P_H, temperature, water activity, electron acceptors, pressure, bioavailability of the pollutants and the presence of nutrients for the success of bioremediation (Leahy and Colwell, 1990). Fertilization of sites using compost, N, P and C has been reported to enhance degradation of pollutants by both the indigeneous microbial communities and the introduced microbes.

Polycyclic Aromatic Hydrocarbons (PAHs) degradation

Many higher-ringed PAHs are resistant toward degradation and described to be toxic, mutagenic or carcinogenic, therefore they are an important risk to the environment and human health. The main input sources of PAH pollutions described by Morgan and Watkinson (1989) are leaches from old storage tanks, road surfaces, domestic waste, oil spills, tanker leakage, incomplete fossil fuel combustion and seepage from natural oil reservoirs. Once these pollutants enter the soil, they are trapped into soil pores and immobilized by adsorption to the soil matrix (Reddy and Sethunathan, 1983). Biological degradation then remains as the most significant means of decomposition. The ability of microbes to degrade oil components was well recognized. The first presentation of the metabolic pathway and the enzymatic reactions resulting in the mineralization of naphthalene was reported by Davies and Evans (1964). Since then, the degradation of PAHs by microbes has been studied intensively and this ability was identified to be present in a wide variety of bacteria, fungi and algae (Leahy and Colwell, 1990). The bacterial PAH catabolic genes often are located on large plasmids together with regulatory genes and in some cases, with genes responsible for chemotaxic toward the PAH (Grimm and Harwood, 1997). The presence of these catabolic plasmids is probably the cause of the adaptation of the indigenous microbes toward PAH pollutants because they are reported to be self-transmissible and horizontal transfer of naphthalene catabolic genes has been described by Stuart-Keil *et al.* (1998).

Many PAHs are very hydrophobic depending upon the number of fused rings and therefore their water solubility is very low. Uptake of PAHs by bacteria proceeds via the water phase and therefore depends upon its solubility in water (Bouwer and Zehinder, 1993). In addition the sorption to soil particles will protect these molecules from degradation by microbes and is the main reason for their long-term persistence in soil. The bioavailability of xenobiotics can be improved by the use of surfactants. Surfactants are amphipathic molecules with both a hydrophobic and a hydrophilic part. Many microbes can produce such surface-active agents referred to as biosurfactant. These accumulate at interfaces which are able to lower the surface tension between interfaces and can form miscelles (Rosenberg, 1993). Intensively investigated organisms and their biosurfactants include *Pseudomonas aeruginosa* strains that produce rhamnolipids (Sim *et al.*, 1997) and *Bacillus* spp. strains that produce surfactins (Fuma *et al.*, 1993). Although the effects of biosurfactants can be dualistic (Allen *et al.*, 1999) and their mode of action and the use of biosurfactants can be considered a valuable tool to improve bioremediation.

Microbial degradation of pesticides

The quest for sustainable agricultural production has resulted in large scale application of pesticides. Ironically, these compounds have also contributed to serious environmental problems arising as a result of their toxicity and persistence in environment. Microbial degradation of pesticides may involve their utilization as sole carbon source or may be through co-metabolic and symbiotic processes, (Swaminathan, 1994). Pesticides which have been primarily developed to control crop pests and weeds in agricultural operations have also made a great impact on human health and preservation of food and food grains.

The effectiveness of biological treatment of pesticides is determined primarily by the type and ability of the microbial population to degrade the pesticide compounds. Microbial action on pesticides is of different types and microorganisms may use the pesticides as sole carbon and energy source. It has been shown that *Pseudomonas* sp. is capable of utilizing several non-chlorinated and mono-chloro substituted analogues of DDT as a sole carbon source (Francis, *et al.*, 1996). Several isolates of *Pseudomonas* from salt mash environment have been found to degrade malathion using it as sole carbon source (Bonguin, 1976) several pesticides have been degraded by co-metabolic process wherein the microorganisms need an external energy source for their extensive breakdown of a pesticide molecule but may cause transformation with reduction, elimination or increase in toxicity. Pesticides such as DDT, 2, 4, 5-T have been subjected to co-metabolic attack by microorganisms (Bollag, 1974).

Pesticides may also be degraded by the symbiotic action of microorganisms. A classic example of this process is the parathion degradation by the symbiotic action of *Pseudomonas stutzeri* and *Pseudomonas aeruginosa* (Alexander, 1972). The process involved non-oxidative and co-metabolic hydrolysis of parathion by *P.stutzeri* the ionic diethyl thiophosphate and p-nitrophenol. *P.aeruginosa* then degrades p-nitrophenol. The diethyl thio-phosphate is exceedingly stable to chemical hydrolysis and microbial attack (Alexander, 1974).

Bollag (1982) reported several mechanisms involved in microbial metabolism which include oxidation, reduction, hydrolysis and dehalogenation of pesticide degradation process. Hydroxylation, dealkylation, cleavage of ether linkages, oxidation of aromatice rings, β-oxidation and epoxidation are the examples of pesticides transformation by oxidative mechanisms (Swaminathan, 1994).

Hydroxylation is a frequent primary reaction of many microorganisms. But the hydroxylation efficiency of species varies depending the site of hydrolysis. For example 2, 4-D is hydroxylated by *Pseudomonas* sp to give 6-hydroxy derivative while *Apergillus* sp gives rise to 2, 4-dichloro-5-hydroxy phenoxy acetic acid and to a lesser extent 2, 5-dichloro-4-hydroxy phenoxy acetic acid (Bollag, 1974). N-alkylation of carbaryl by *Aspergillus terreus* gives 1-naphthyl-N-hydroxy methyl carbamate and subsequently 1-naphthyl carbamate. Hydrolytic mechanism have been reported in the breakdown of carbamate and organophosphorus pesticides. Malathion is hydrolytically cleaved to phospharathate compound and carboxylic acid derivatives by *Pseudomonas* sp. and *Trichoderma viride* (Bollag, 1974).

It has been shown that *Hydrogenomonas* sp. is able to cleave one of the rings of DDT aerobically provided as it is first allowed to metabolic the substrate anaerobically. In this process it is found that a single organisms is able to convert DDT to a product, p-chlorophenyl acetic acid. This product is an excellent substrate for a variety of microorganisms. It has been demonstrated that DDT is essentially completely degraded by microorganisms (Swaminathan and Shivaraman, 1980).

It is believed that a wide spectrum of pesticide degradation plasmids may already have evolved in response to the widespread use of a variety of complex synthetic pesticide molecule. The exchange of plasmids in a microbial population may be an important factor in increasing the range of degradative activities of microorganisms. The fate of pesticides in environment represents an area of major significance determining their degree of usefulness and potential hazards (Swaminathan, 1994).

Conclusion

The use of chemical fertilizers and pesticides at high level for intensive agriculture system for increasing crop production is not only deleterious to the environment by affecting the soil system but injurious to human health. Microorganisms have great potential of degrading pesticides used in agricultural practices and microbes are the potential biological system in maintaining long term soil fertility and sustainable soil health management. Microorganisms are dynamic biological agents plays significant role in decomposing the organic mateials on earth and also widely exploited to clean up the environmental problems in nature. Bioremediation process is the action of microbial system to degrade environmental problems and it is being used in *insitu* without the removal and transport of polluted soil and without the disturbance of the soil matrix. It is a suitable biological tool to restore contaminated sites. The selection and exploitation of suitable soil and rhizosphere bacteria or fungi or actinomycetes or communities of organism would be able to sustain and proliferate on the root system of a plant which is suitable for rhizoremediation. Construction of genetically modified strains is particularly more relevant for the degradation of pollutants and the genetically engineered *Pseudomonas fluorescens*—HK-44 which degrades the naphthalene.

References

Allen, C.C.R., Boyd, D.R., Hempenstall, F., Larkin, M.J. and Sharma, R. 1999. Contrasting effects on nonionic surfactant on the biotransformation of polycyclic aromatic hydrocarbons to cis-dehydrodiols by oil bacteria. *Appl. Environ. Microbiol.*, **65**: 1335-1339.

Alexander, M. 1972. Environmental Toxicology of Pesticides. Masumara, F. (Ed.), Academic Press, UK.

Alexander, M. 1974. Pesticide degradation by symbiotic action of microorganisms. *Adv. Appl. Microbiol.*, **18**: 1.

Anderson, T.A., Guthrie, E.A. and Walton, B.T. 1993. Bioremediation in the rhizosphere. *Environ. Sci. Technology*, **27**: 2630-2636.

Aprill, W. and Sims, R.C. 1990. Evaluation of the use of prairie grasses for stimulating polycyclic aromatic hydrocarbon treatment in soil. *Chemosphere*, **20**: 253-265.

Babich, H. and Stotzky, R. 1978. Heavy metals in natural aquatic and terrestic systems. In: *Aquatic*

Toxicology (Ed.) Nariagu, J. McGrow Hill Publication, New York, USA.

Bollag, J.M. 1974. Degradation of 2, 4-D by *Pseudomonas* and *Aspergillus. Adv. Appl. Microboil.*, **18**: 75.

Bonguin, A.W. 1976. Degradation of malathion. *Appl. Environ. Microbiol.*, **33**: 356.

Bouwer, E.J. and Zehinder, A.J.B. 1993. Bioremediation of organic compounds putting microbial metabolism to work. *Tibtech.*, **11**: 360-367.

Brazil, G.M., Kenefik, L., Callana, L.M., Haro, A., de Lorenzo, V., Dowling, D.N. and Gara, F.O. 1995. Construction of rhizosphere Pseudomonad with potential to degrade polychlorinated biphenyls and detection of bph gene expression in the rhizosphere. *Appl. Environ. Microbiol.*, **61**: 1946-1952.

Dean-Ross, D. 1987. Biodegradation of toxic wastes in soil. *ASM* News, **53**: 490-492.

Dixon, B. 1996. Bioremediation is here to stay. *ASM* News, **62**: 527-528.

Due, M., Sethunathan, N. and Johni, A.K. 2002. Biotechnology and bioremediation successes and limitations. *Appl. Microbiol. Biotechnol.*, **59**: 143-152.

Ford, C.Z., Sayler, G.S. and Burlage, R.S. 1999. Contaminent of genetically engineered microorganism during a field bioremediation application. *Appl. Microbiol. Biotechnol.*, **51**: 397-400.

Francis, M.S., Philip, B. and George, P.J. 1996. Extraction of copper abundant metallothionein from *Synechocystis Salina. Inte. Sympo Cyanobacterial Biotechnol.*, Trichy, Tamil Nadu.

Fuma, S., Fujishima, Y., Corbell, N., Souza, C., Nakano, M.M., Zuber, P. and Yamane, K. 1993. Nucleotide sequence of 5 portion of SrFA that contains the region required for competence establishment in *Bacillus subtilis, Nucleic Acids Res.*, **21**: 93-97.

Grimm, A.C. and Harood, C.S. 1997. Chemotaxis of *Pseudomonas* spp. to the polyaromatic hydrocarbon naphthalene. *App. Environ. Microbiol.*, **63**: 4111-4115.

Goldsteing, R.M., Mallor, L.N. and Alexander, M. 1985. Reasons for possible failure of inoculation to enhance biodegradation. *Appl. Environ. Microbiol.*, **50**: 977-983.

Heidelberg, J.F., Paulsen, I.T., Nelson, K.I., Gaidos, E.J., Nelson, W.C., Rea, T.D. and Eisen, J.A. 2002. Genomic sequence of the dissimilatory metal iron–reducing bacterium, *Shewanella oneidensis, Nature Biotechnol.*, **2**: 1-6.

Kannaiyan, S. 2002. Biofertilizers for rice production. In: *Biotechnology of Biofertilizers.* Kannaiyan, S. (Ed.) pp. 9-28, Narosa Publishing House, New Delhi.

King, J.M.H., Digrazia, P.M., Applegate, B., Lavimer, F. and Sayler, G.S. 1990. Rapid, sensitive bioluminescent reporter technology for naphthalene exposure and biodegradation, *Science*, **249**: 778-781.

Kuiper, I., Lagendik, E.L., Bloemberg, G.V. and Lugtenberg, B.J.J. 2004. Rhizoremediation, A. Beneficial Plant—Microbe interaction. *Mol. Plant Microbe Interaction.*, **17**(1): 6-15.

Leahy, J.G. and Colwell, R.R. 1990. Microbial degradation of hydrocarbons in the environment. *Microbiol Rev.*, **54**: 305-315.

Morgan, P. and Watkinson, R.J. 1989. Hydrocarbon degradation in soil and methods for soil biotreatment. *Crit. Rev. Biotechnol.*, **8**: 305-333.

Palaniappan, S.P. and Annadurai, K. 1999. *Organic Farming: Theory and Practice*, Scientific Publishers (India), Jodhpur.

Reddy, B.R. and Sethunathan, N. 1983. Mineralization of parathion in the rice rhizosphere. *Appl. Environ Microbiol.*, **45**: 826-829.

Rosenberg, E. 1993. Exploiting microbial growth on hydrocarbons—new markets. *Tibtech*, **11**: 419-424.

Sayler, G.S. and Ripp, S. 2000. Field application of genetically engineered microorganisms for bioremediation process. *Curr. Opin. Biotech.*, **11**: 286-289.

Sim, L., Ward, O.P. and Li, Z.Y. 1997. Production and characterization of biosurfactant isolated from *Pseudomonas aeruginosa* UW-1. *J. Inde. Microbiol. Biotechnol.*, **19**: 232-238.

Stuart-Keil, K.G., Hohnstock, A.M., Herrick, J.B. and Madsen, E.L. 1998. Plasmids responsible for horizontal transfer of naphthalene catabolism genes between bacteria at a coal tar contaminated site are homologous to PDTGI from *Pseudomonas putida* NCIB-9816-4. *Appl. Environ. Microbiol.*, **64**: 3633-3640.

Swaminathan, T. 1994. Microbial degradation of pesticides wastes. In: *Microbes for better living.* Sankaran, R. and Manja, K.S. (Eds.). The Bangalore Printing and Publishing Company Ltd., Bangalore.

Swaminathan, T. and Shivaraman, N. 1981. Degradation of chlorinated hydrocarbon pesticides. *Proc. Natl. Workshop on microbial degradation of industrial wastes*, Nagpur.

Volesky, B. 1987. Toxicity of heavy metals and recovery from aquatic systems. *Trends in Biotechnology,* **5**: 96-101.